First edition: First printing

ISBN-13:978-1530623693

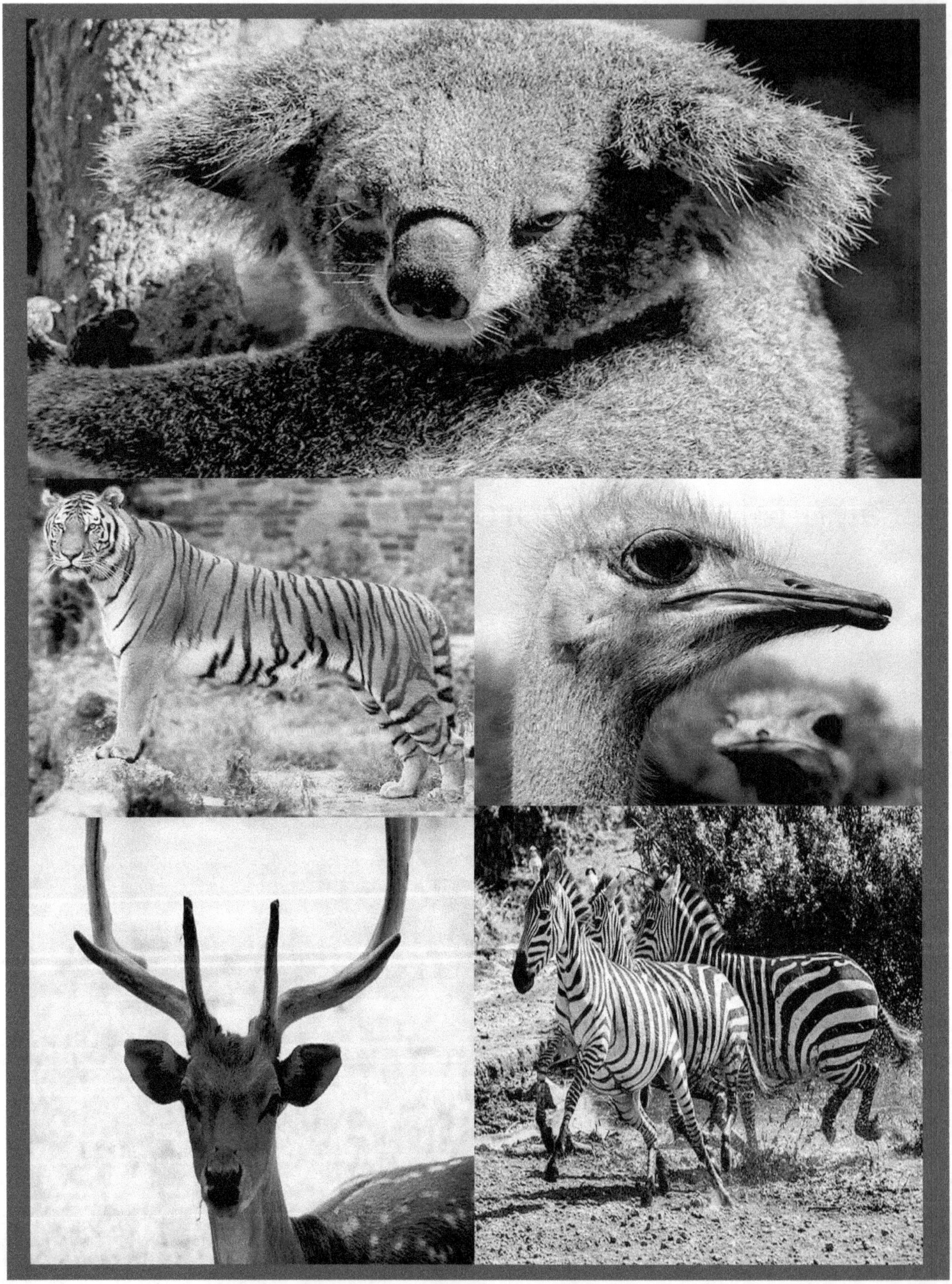

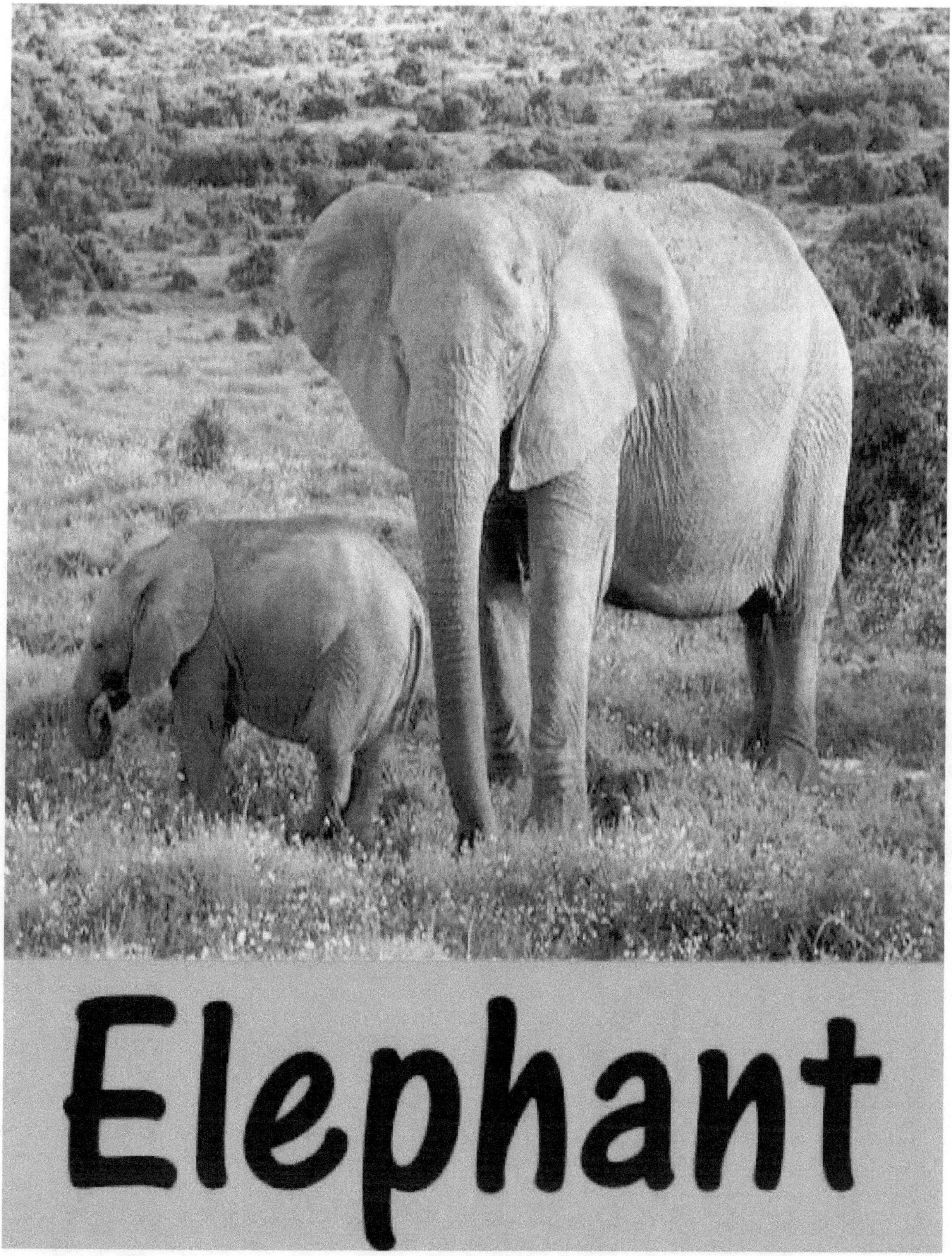

Elephant

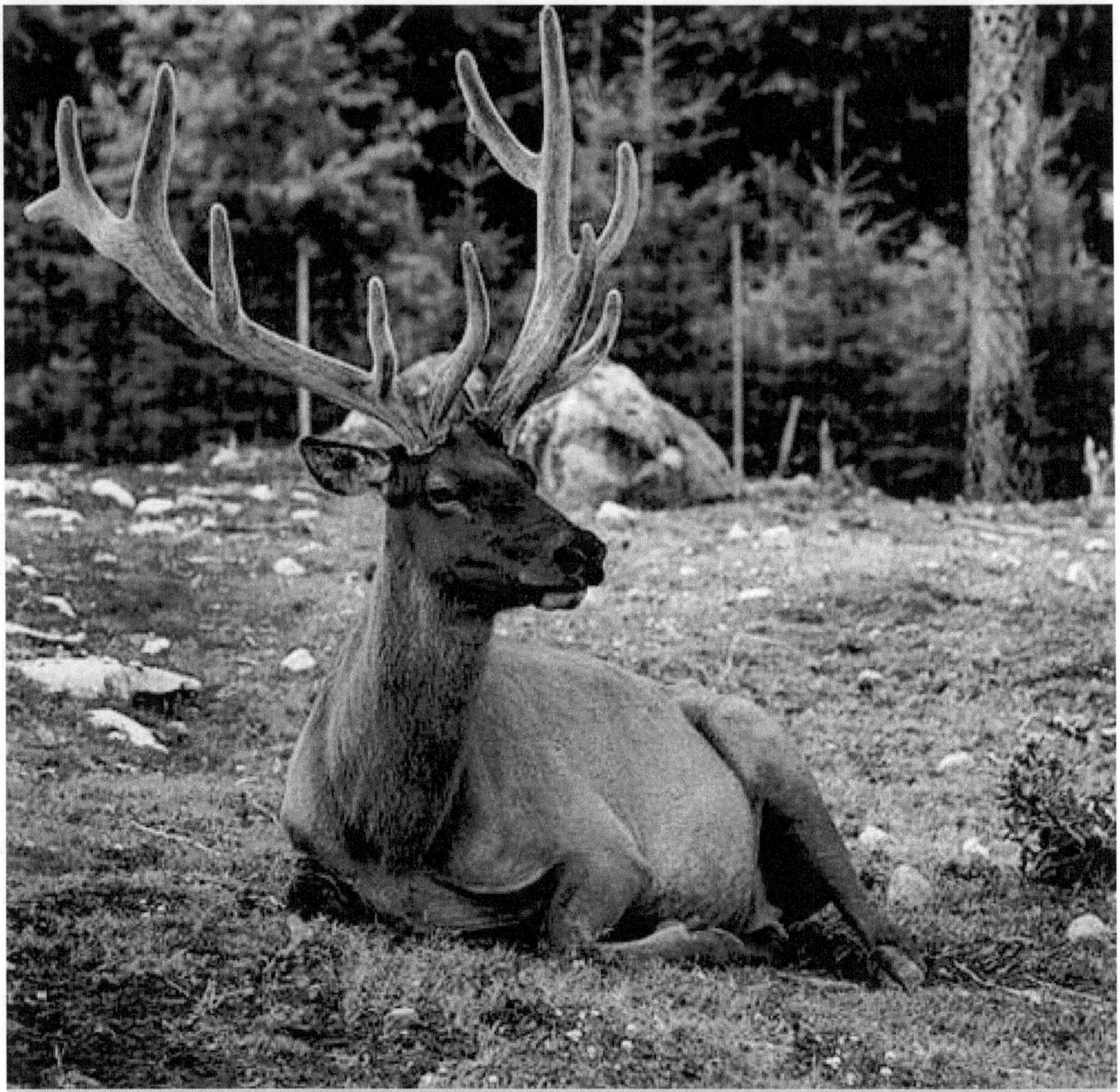

Deer

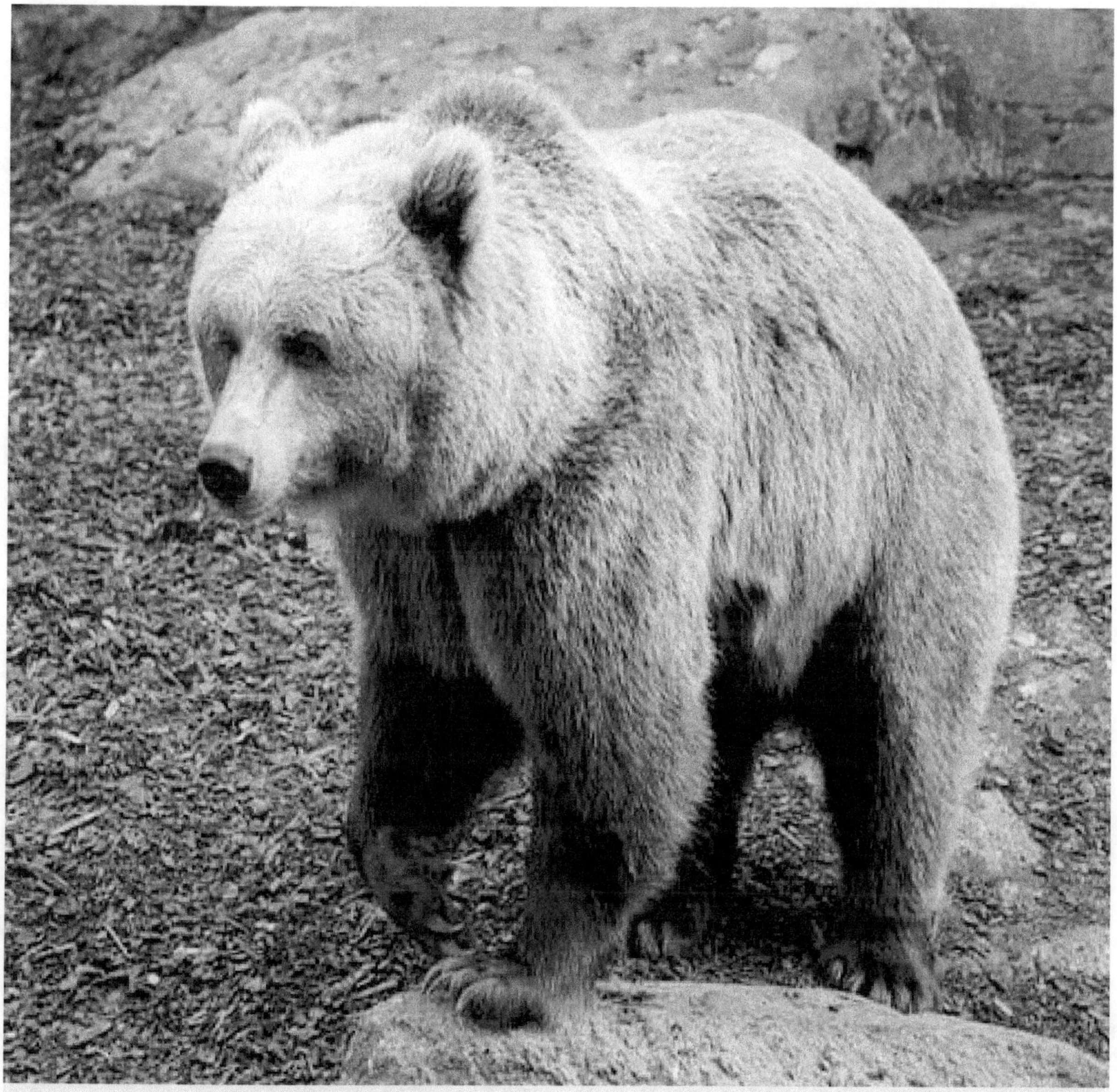

Bear

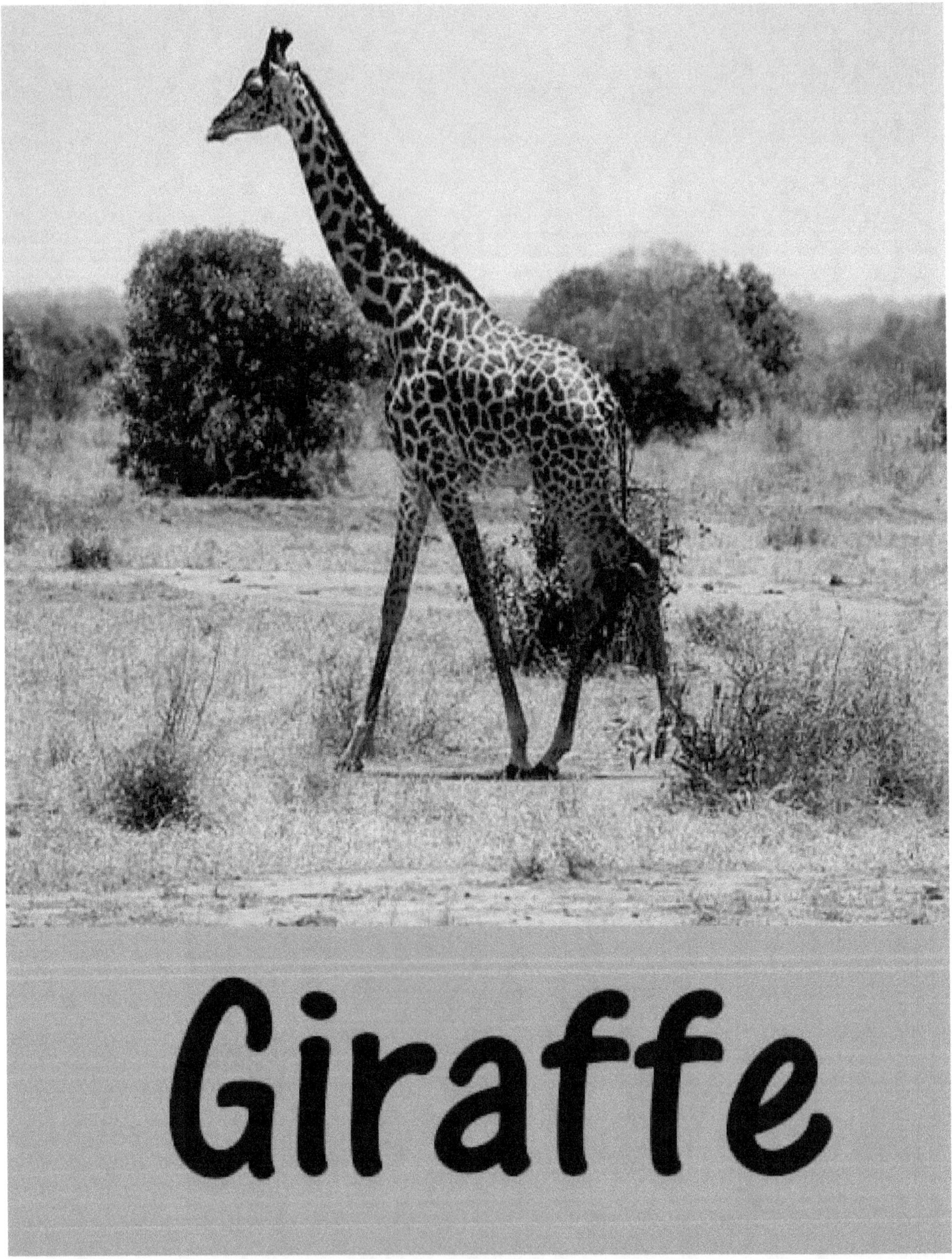

Giraffe

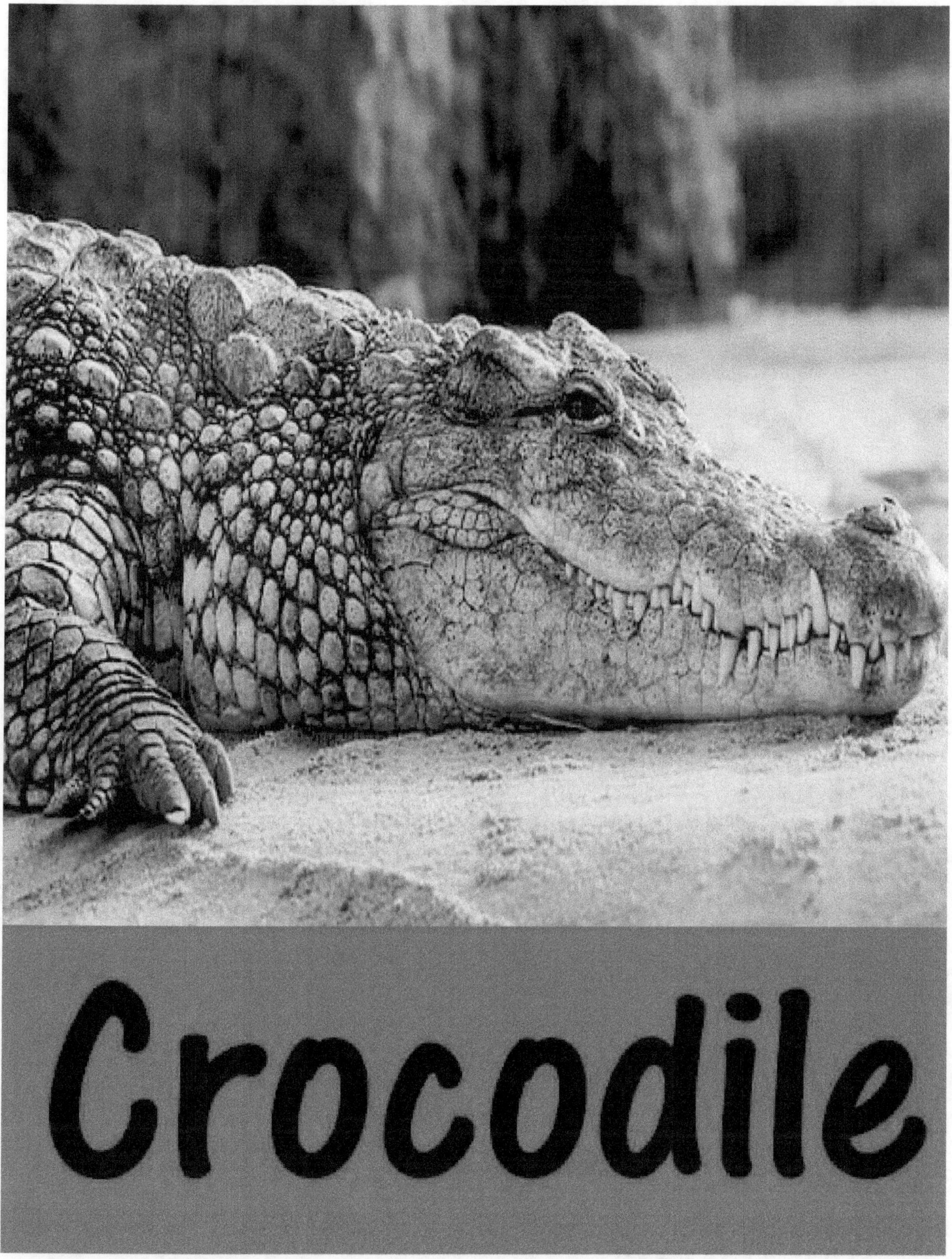

Crocodile

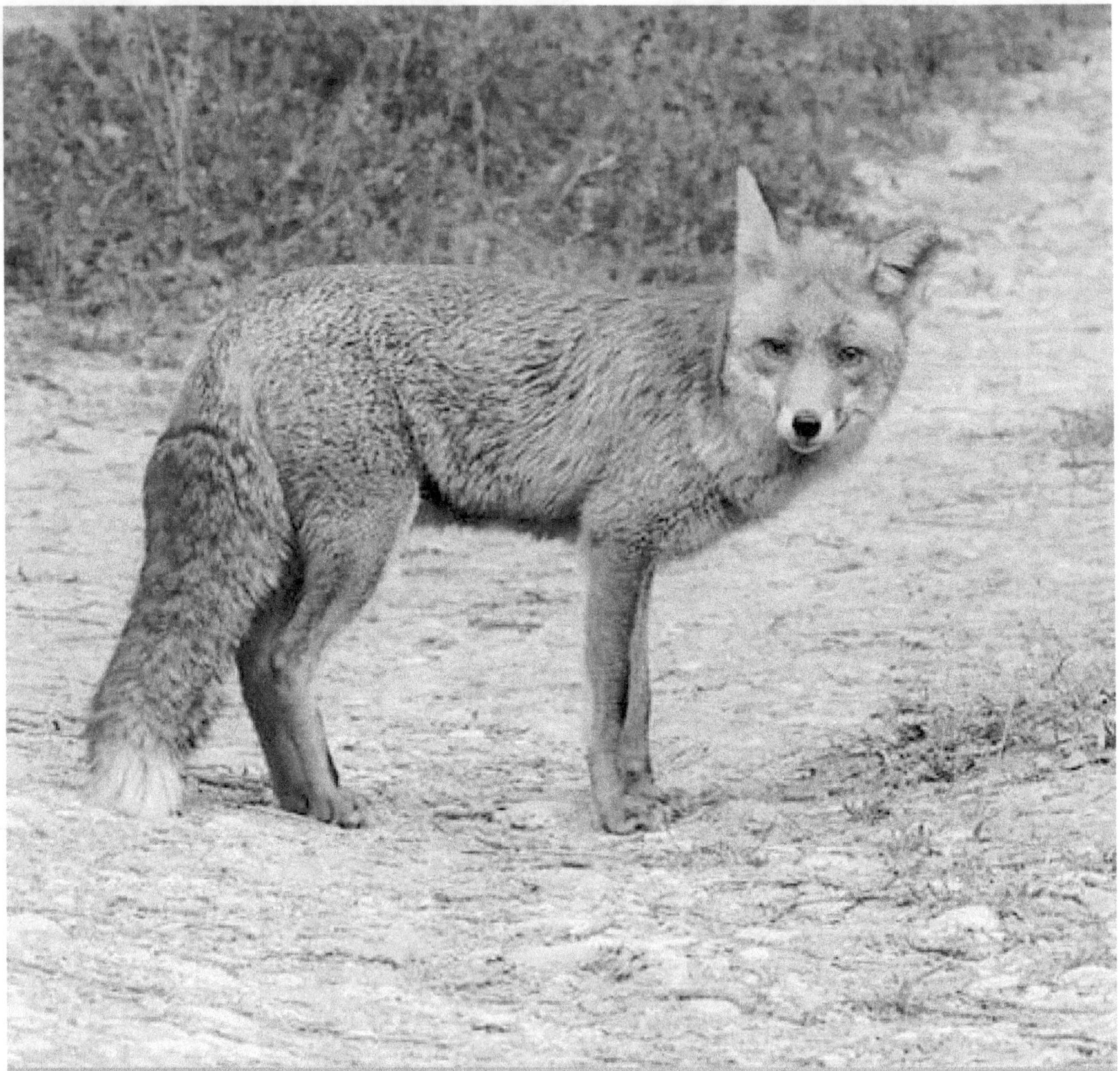

Fox

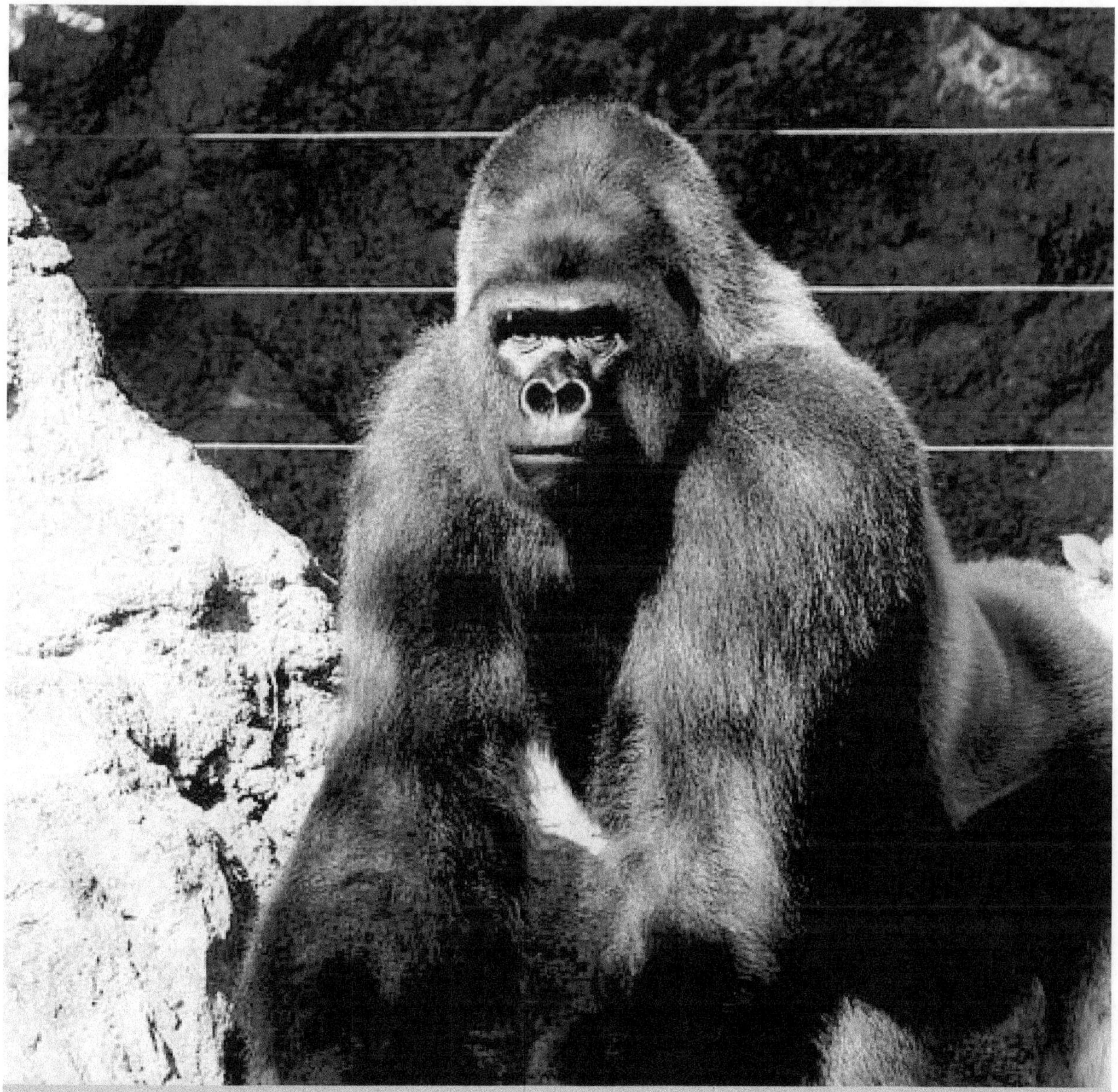

Gorilla

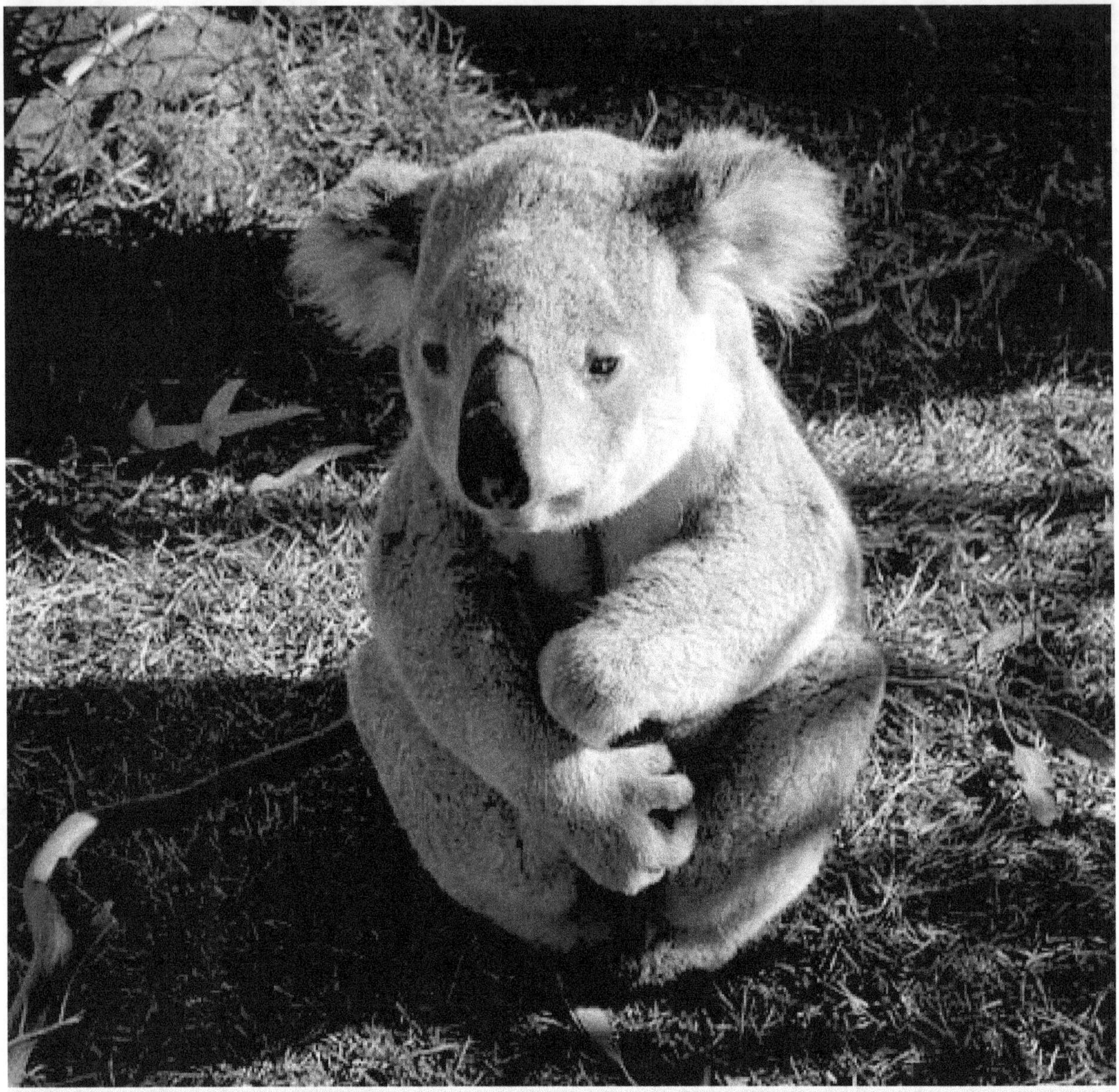

Koala

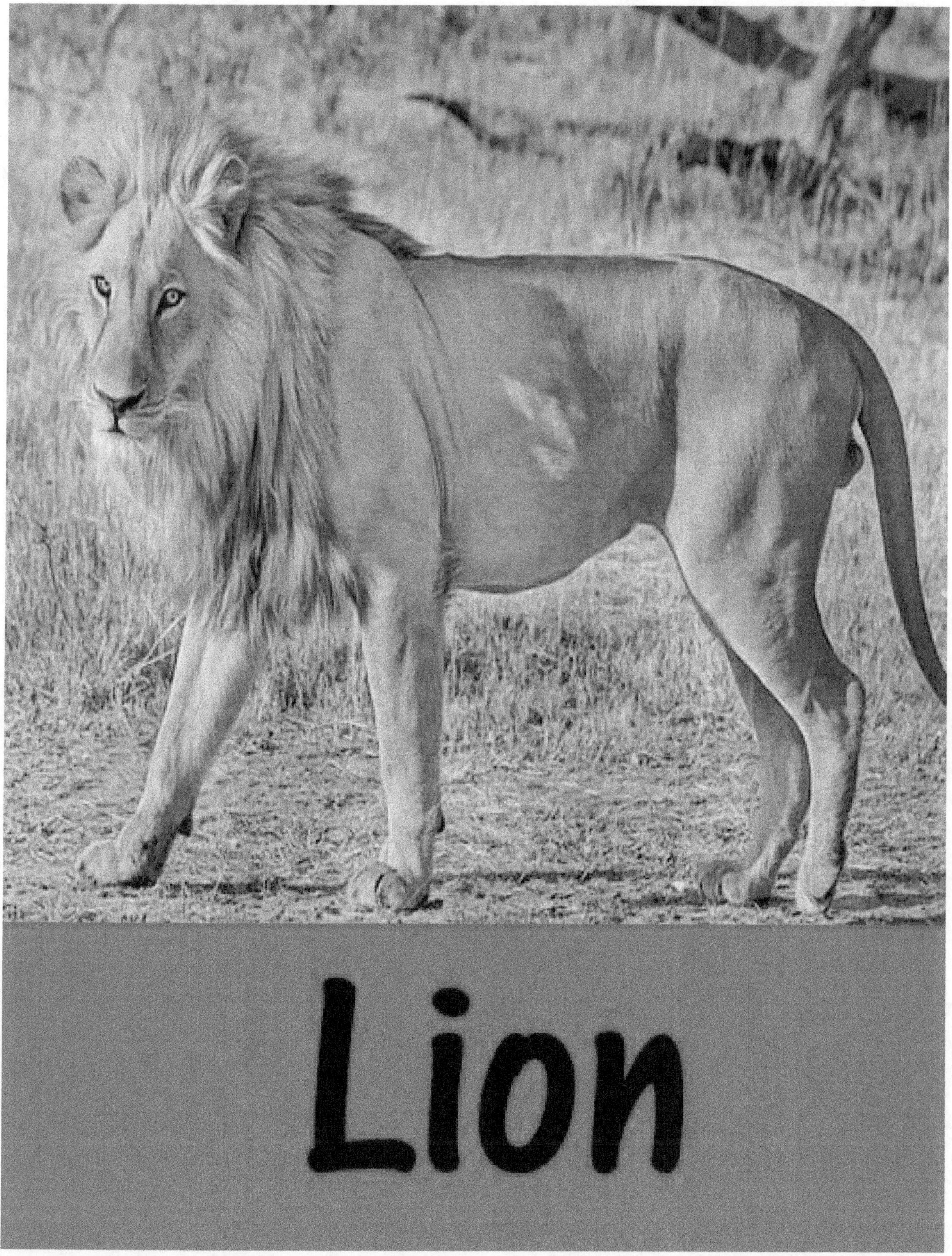

Lion

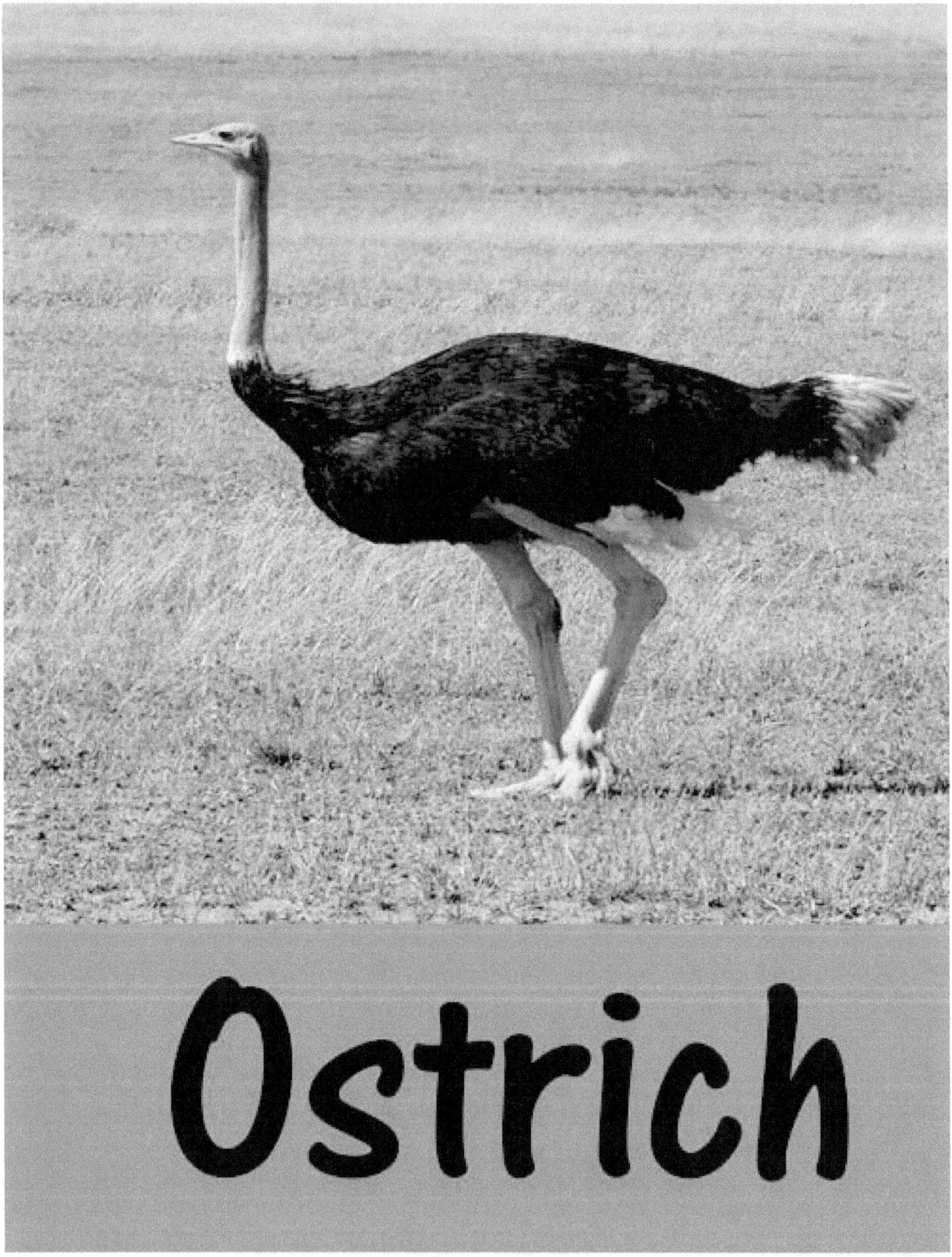

Ostrich

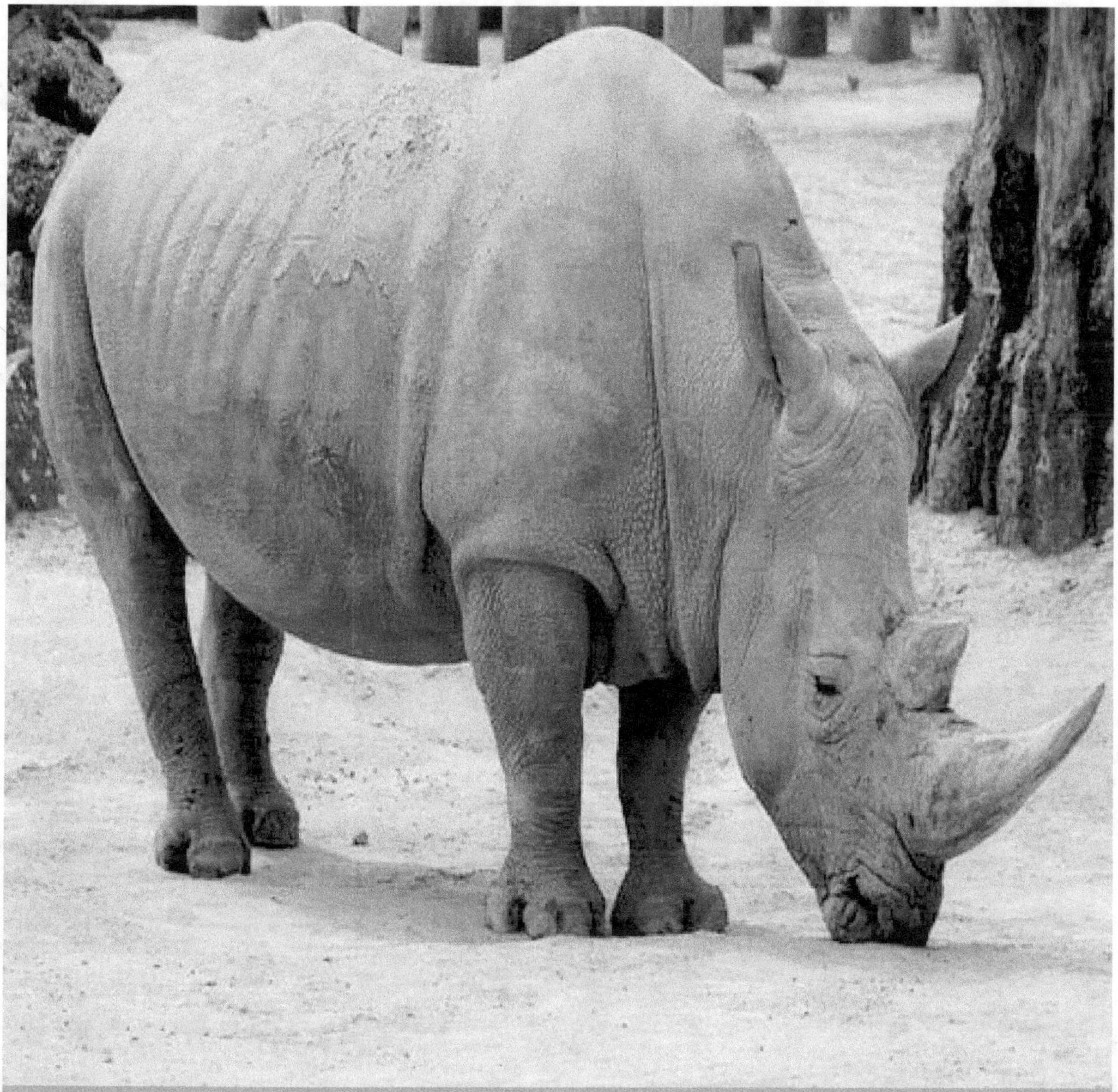

Rhinoceros

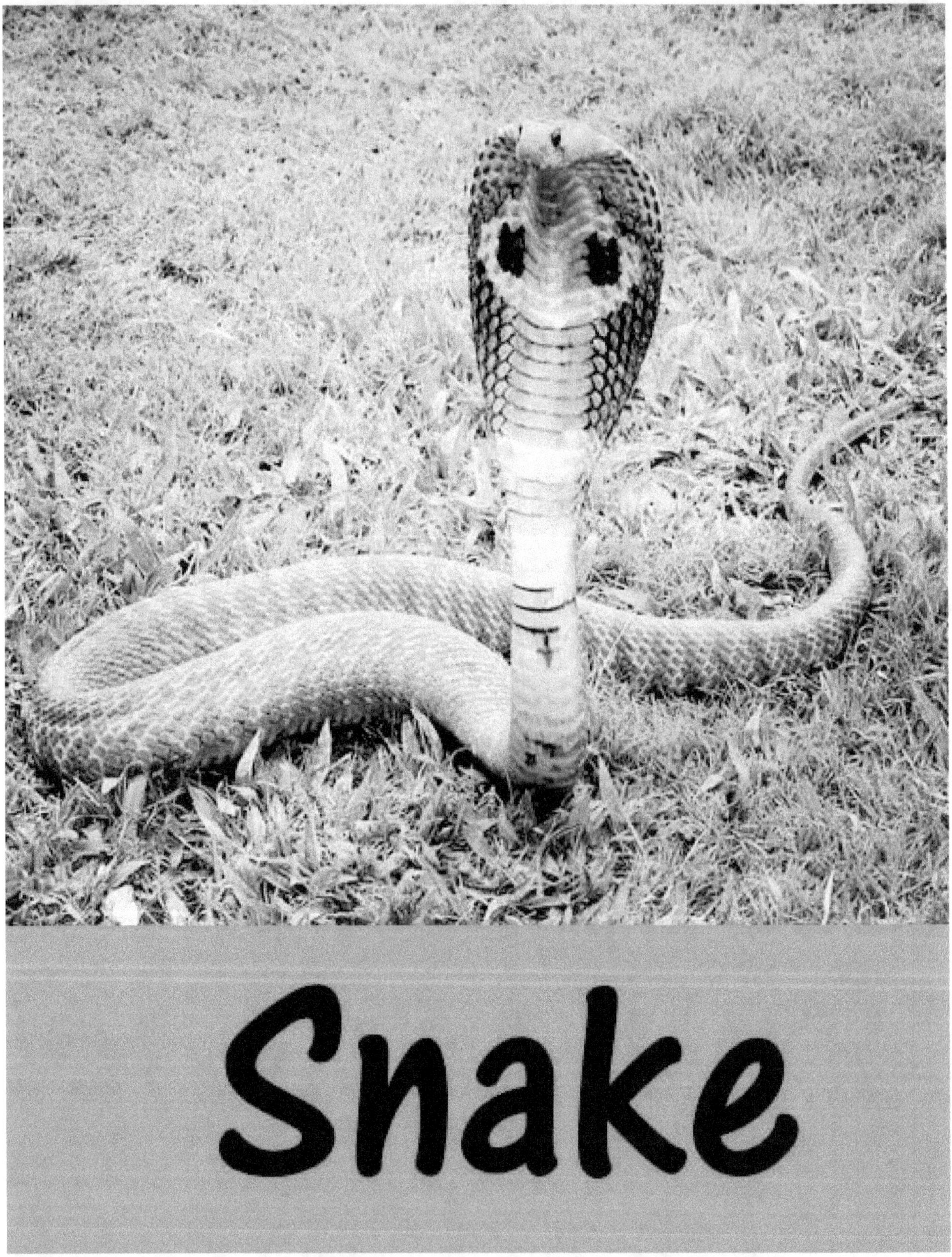

Snake

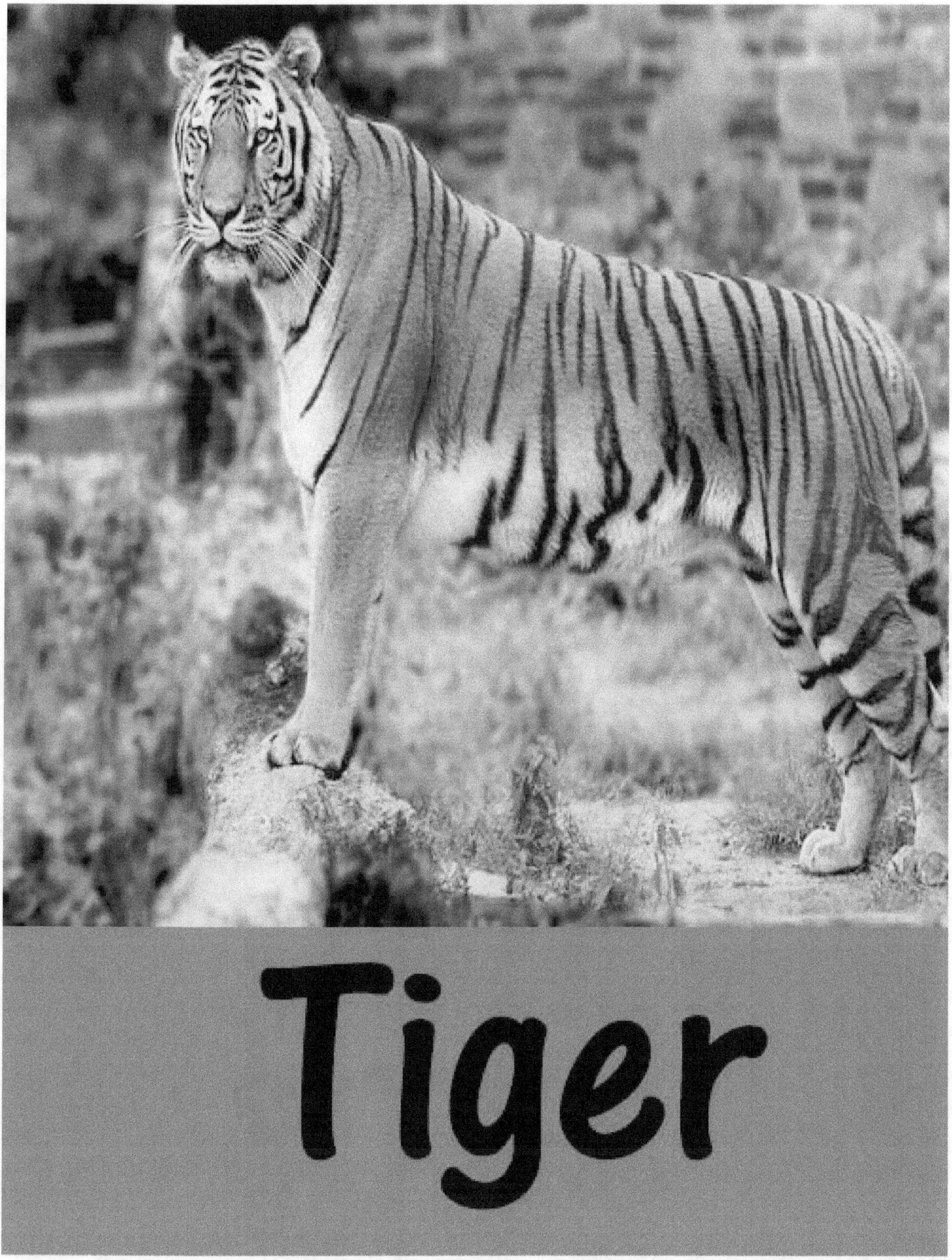

Tiger

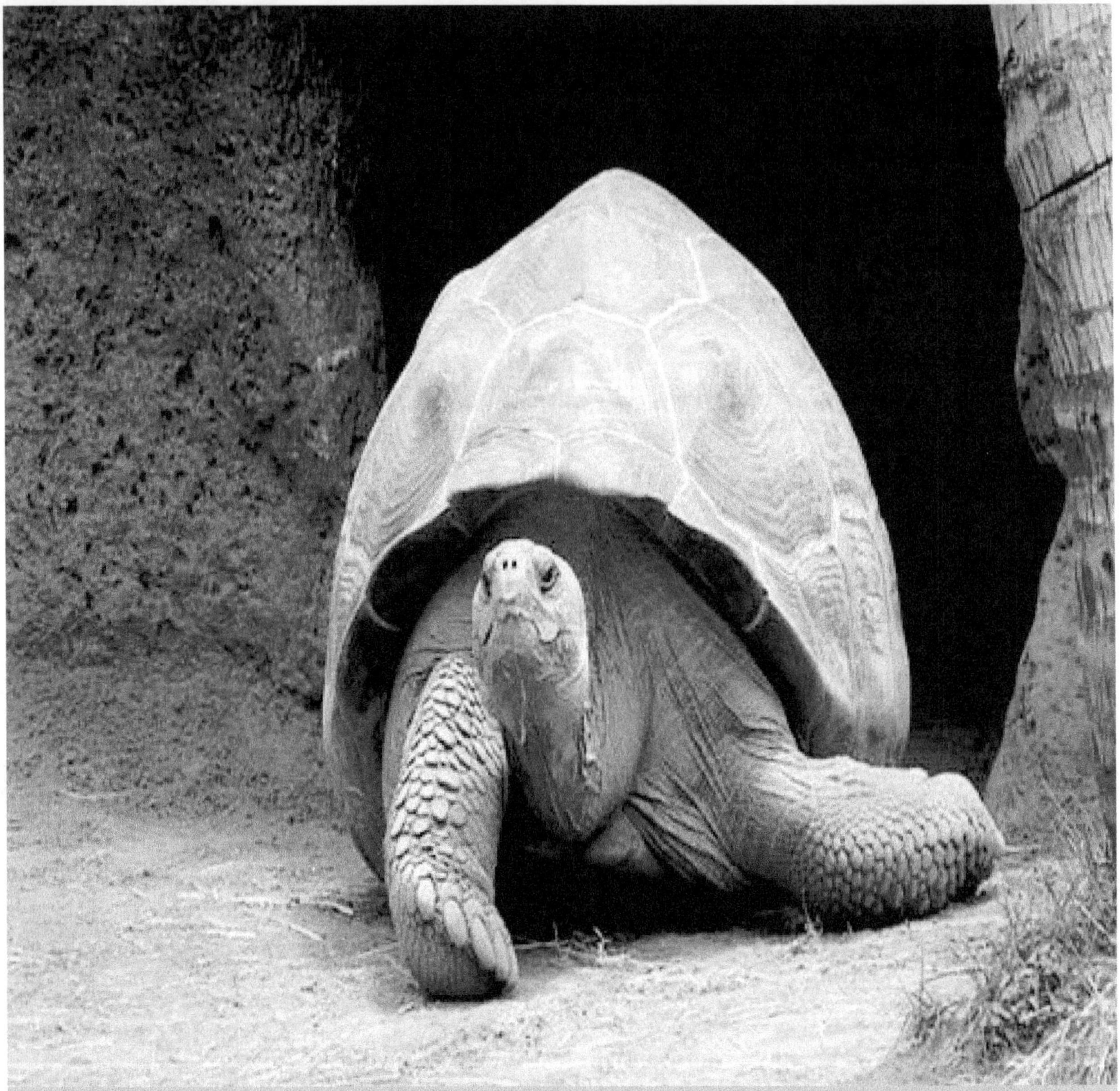

Turtle

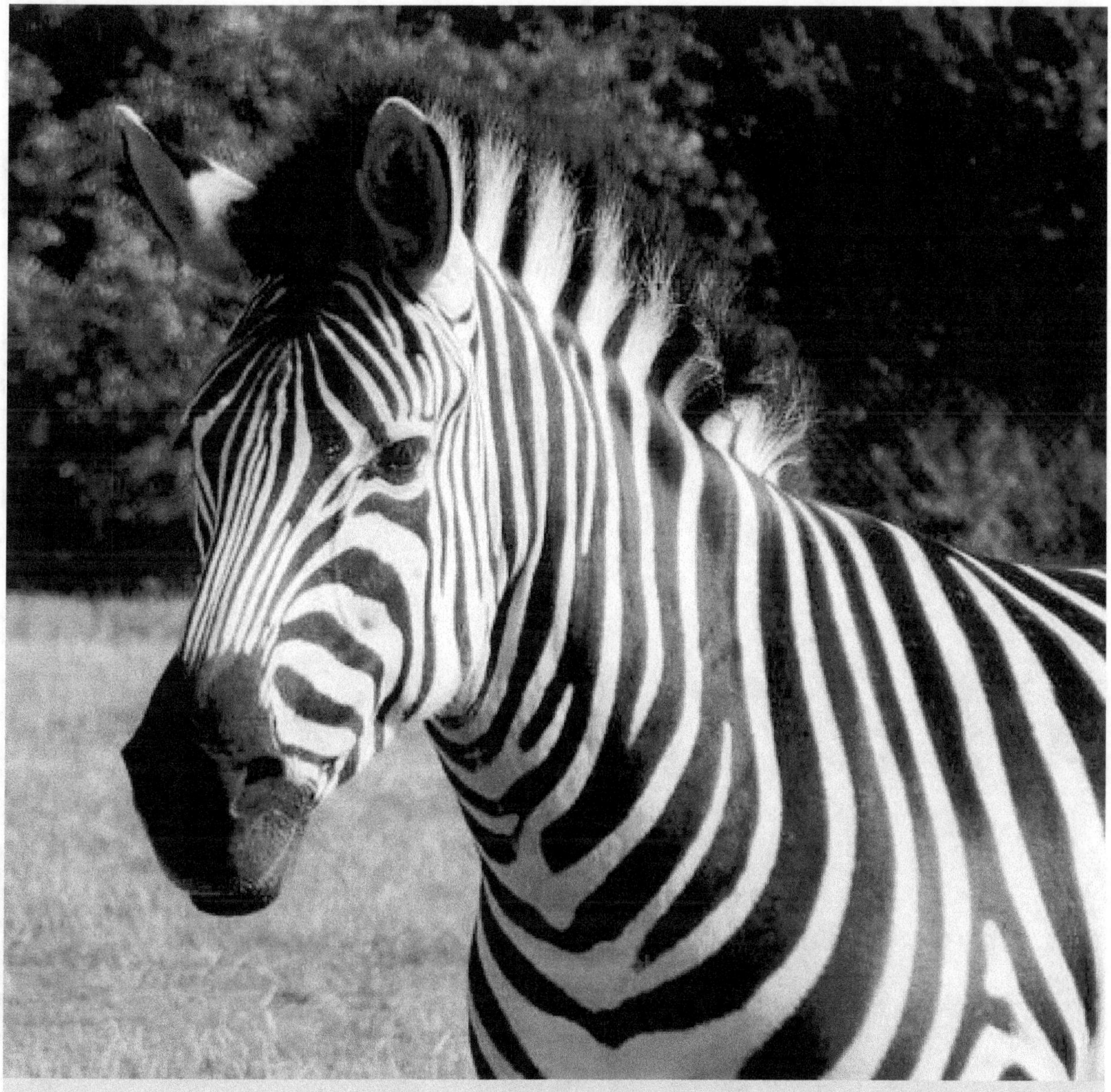

Zebra

Do you remember me?

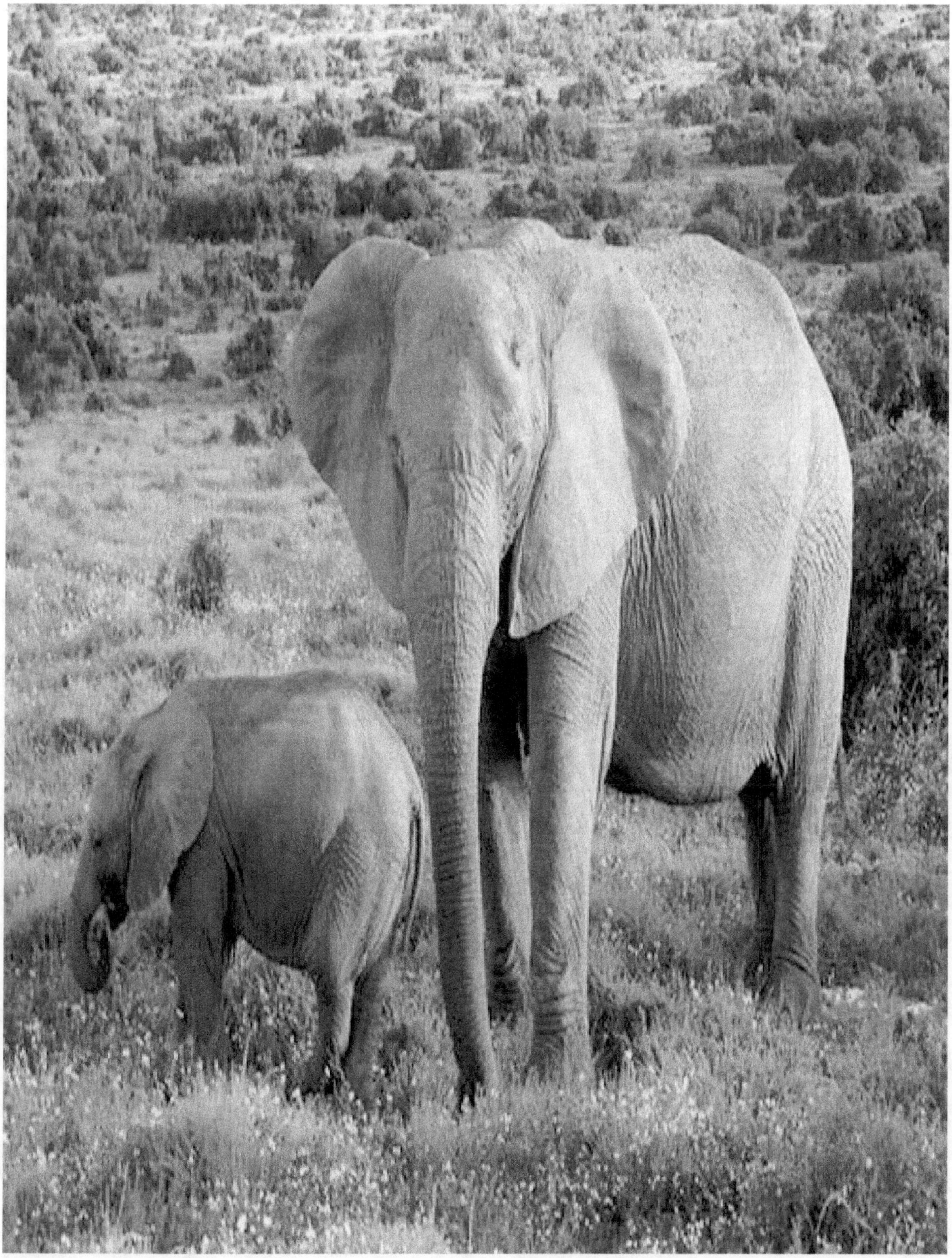

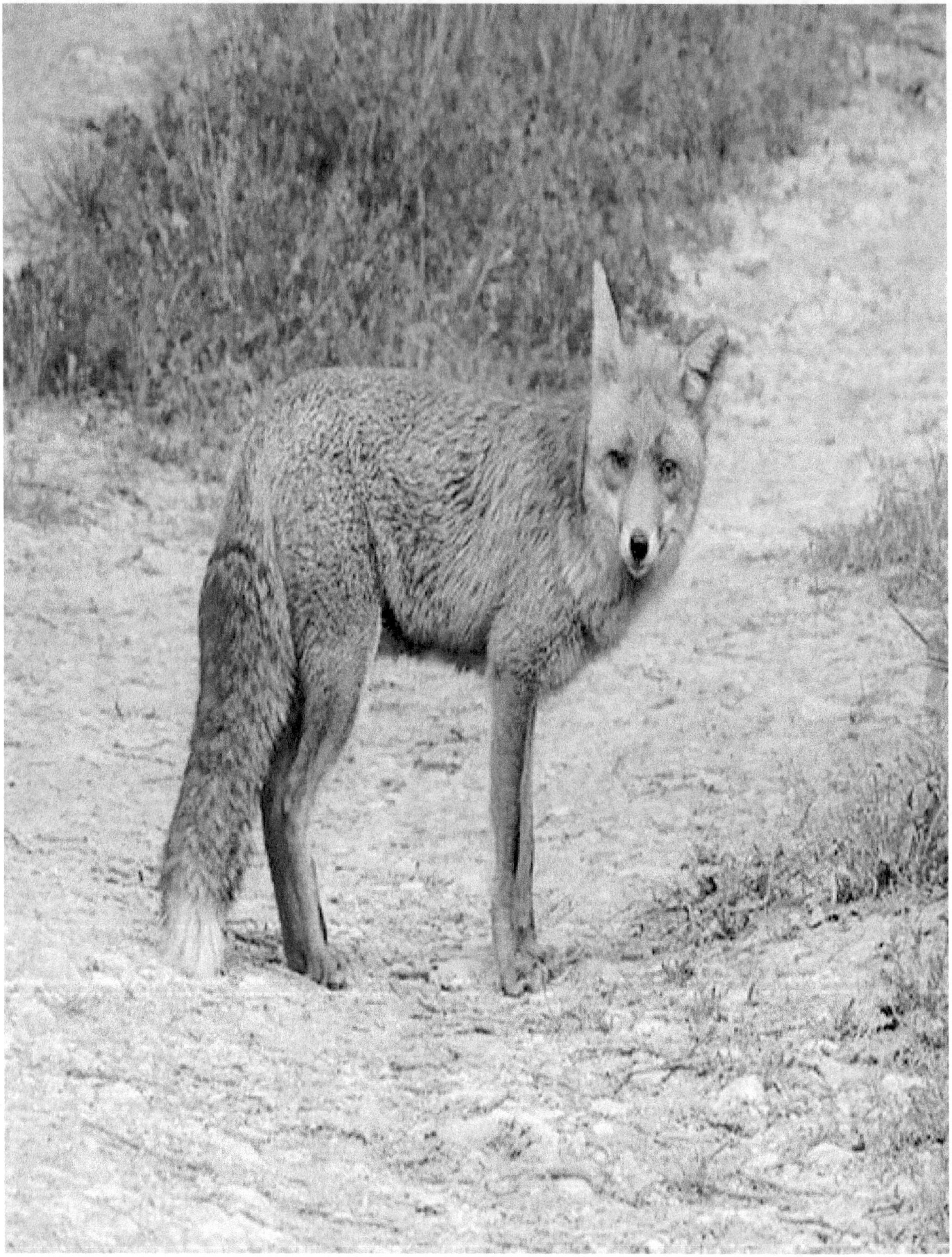

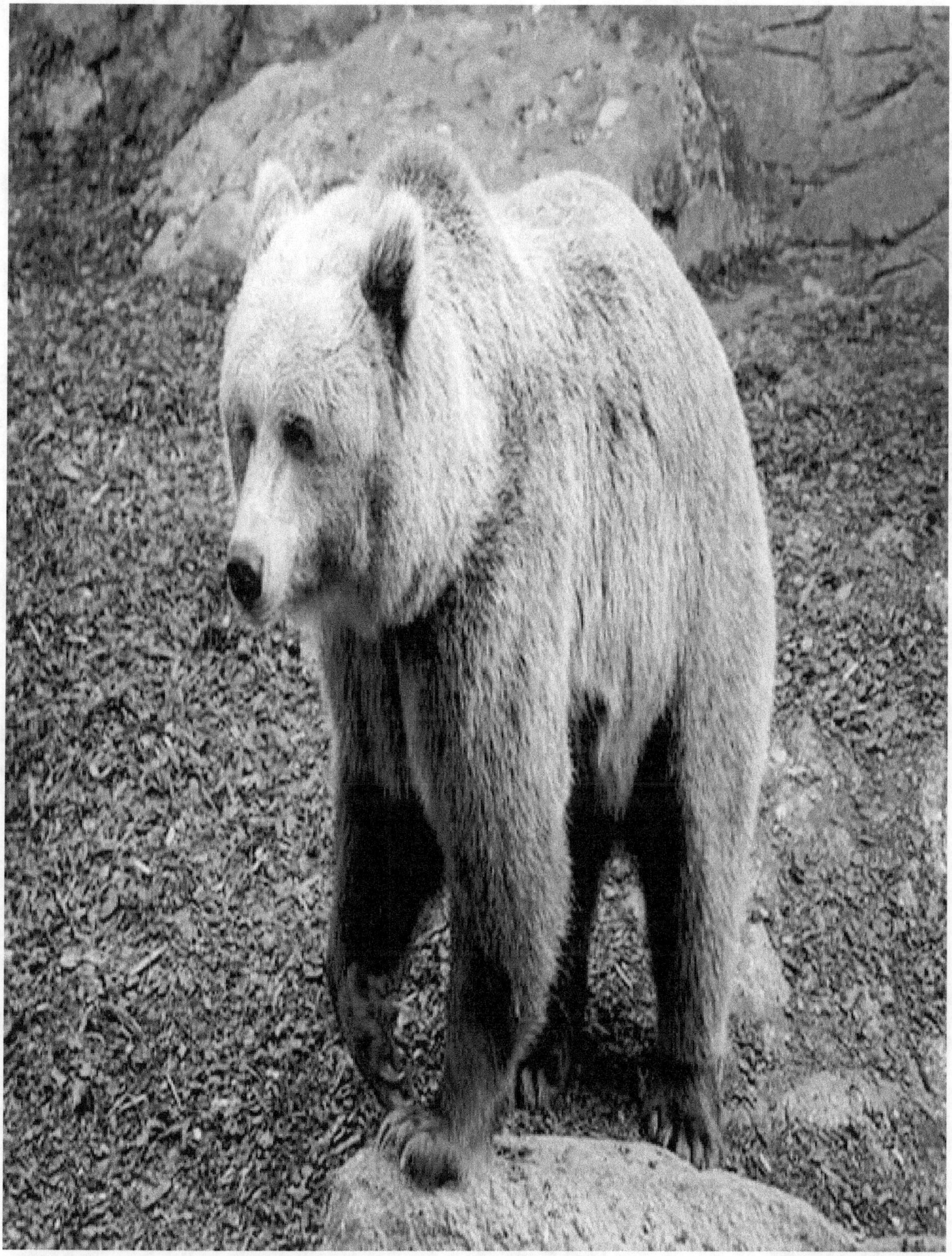

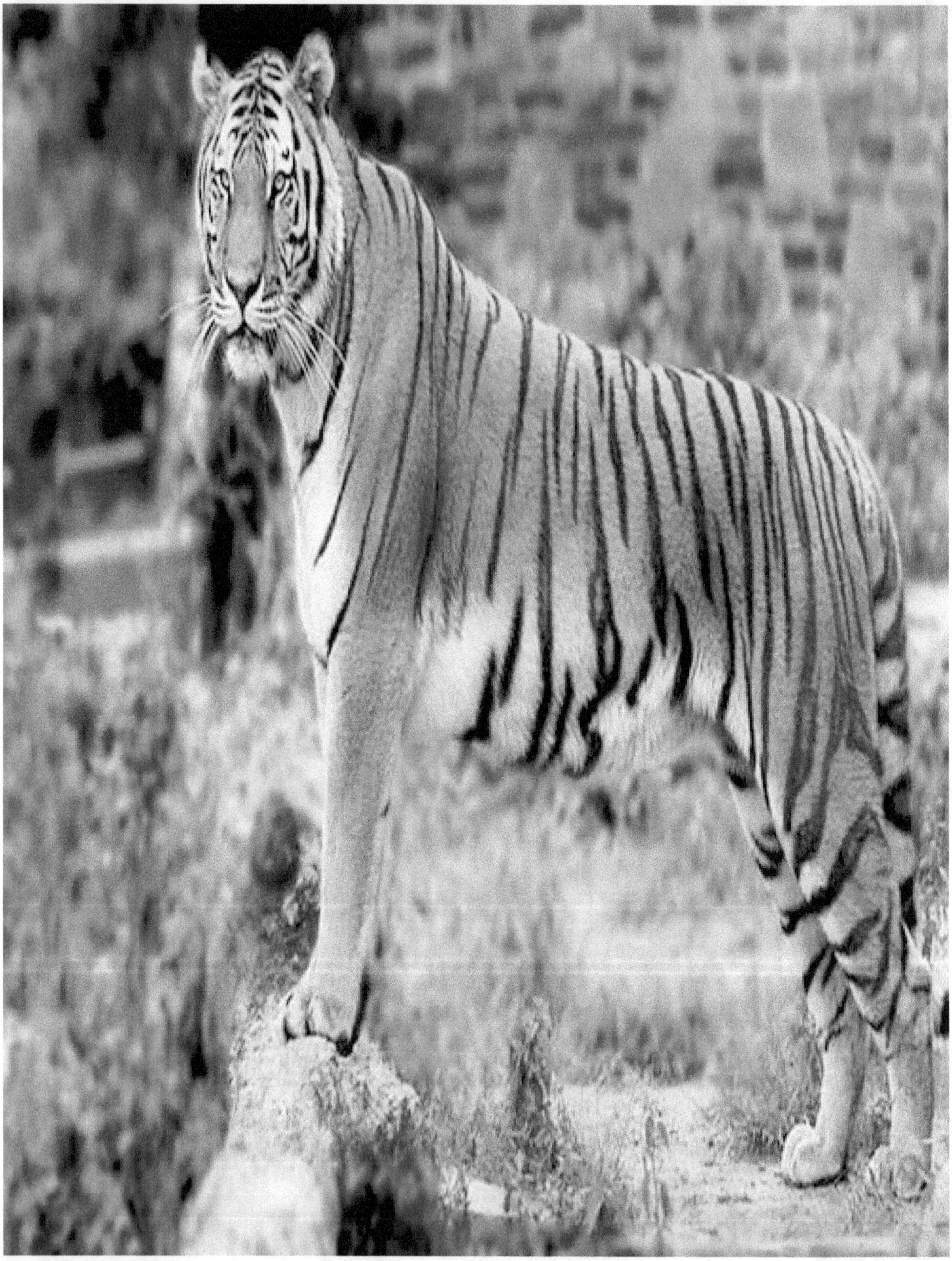

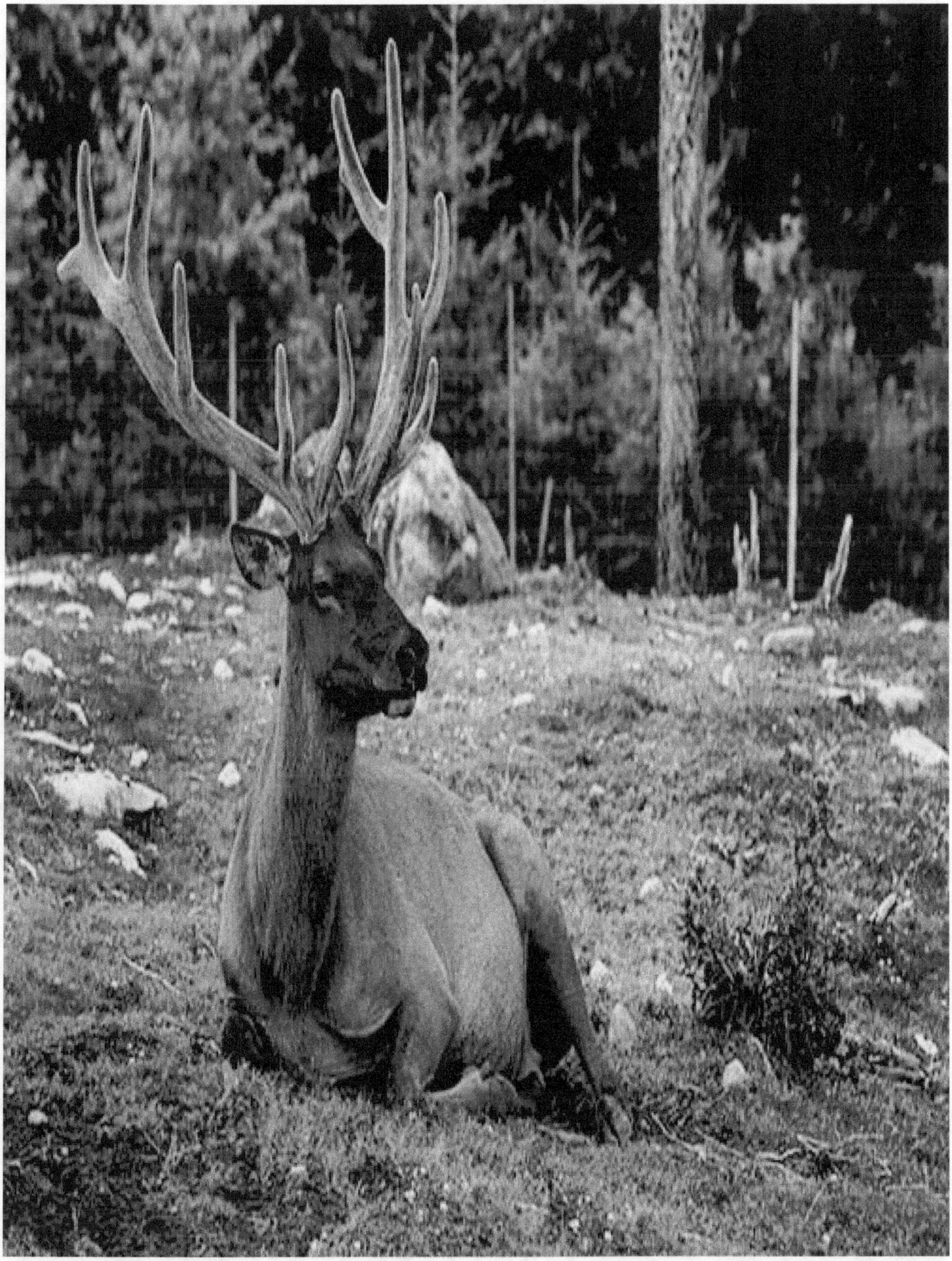

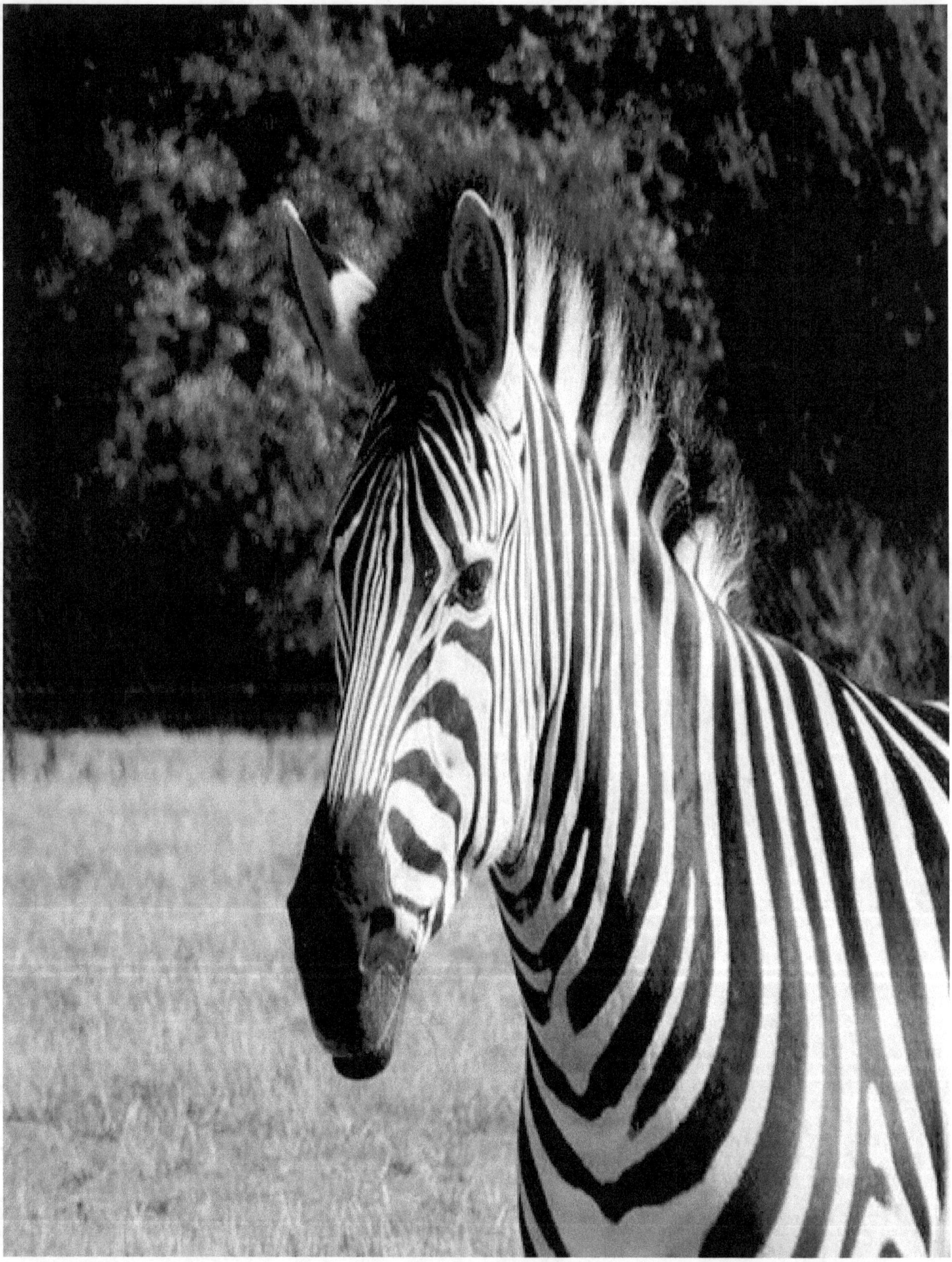

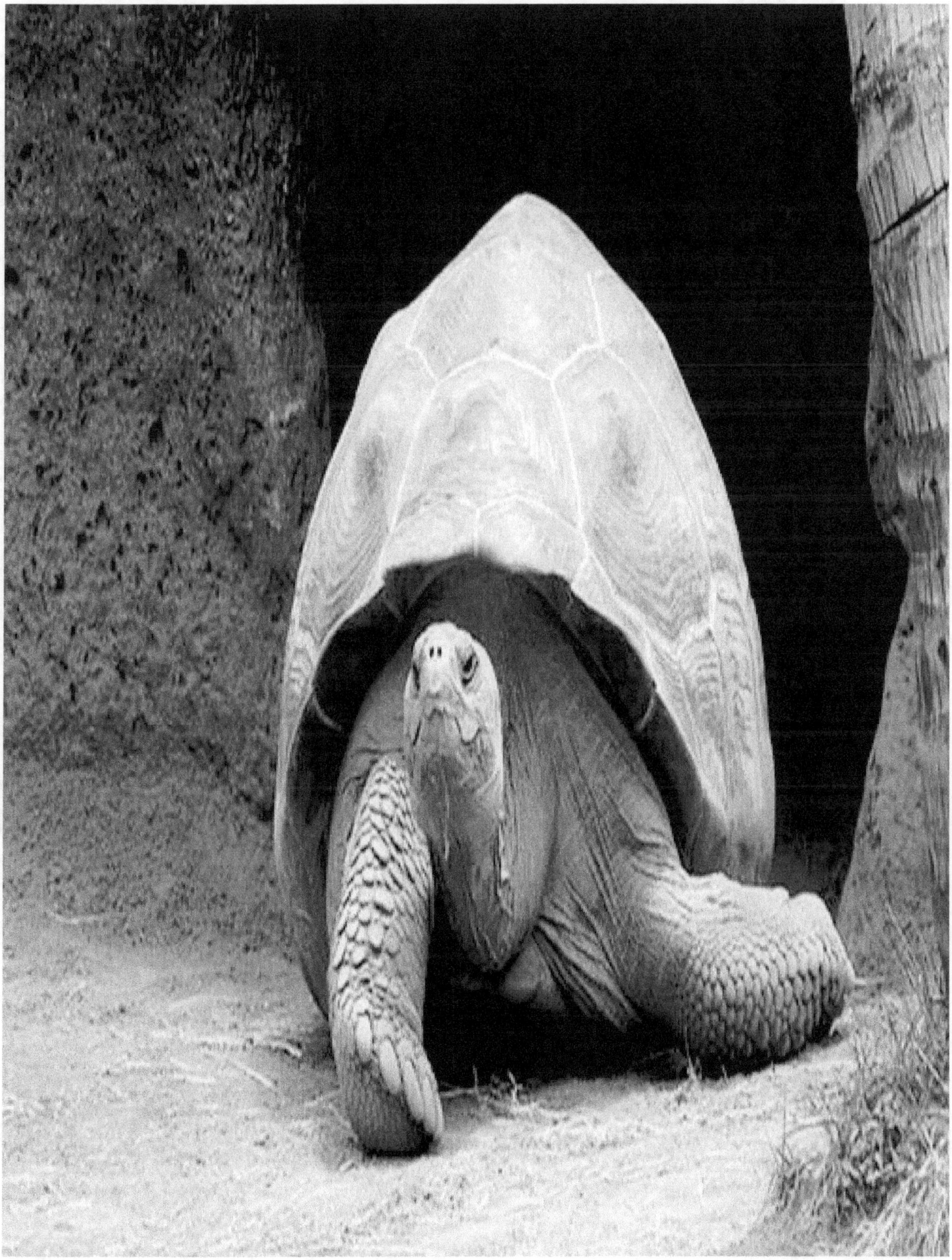

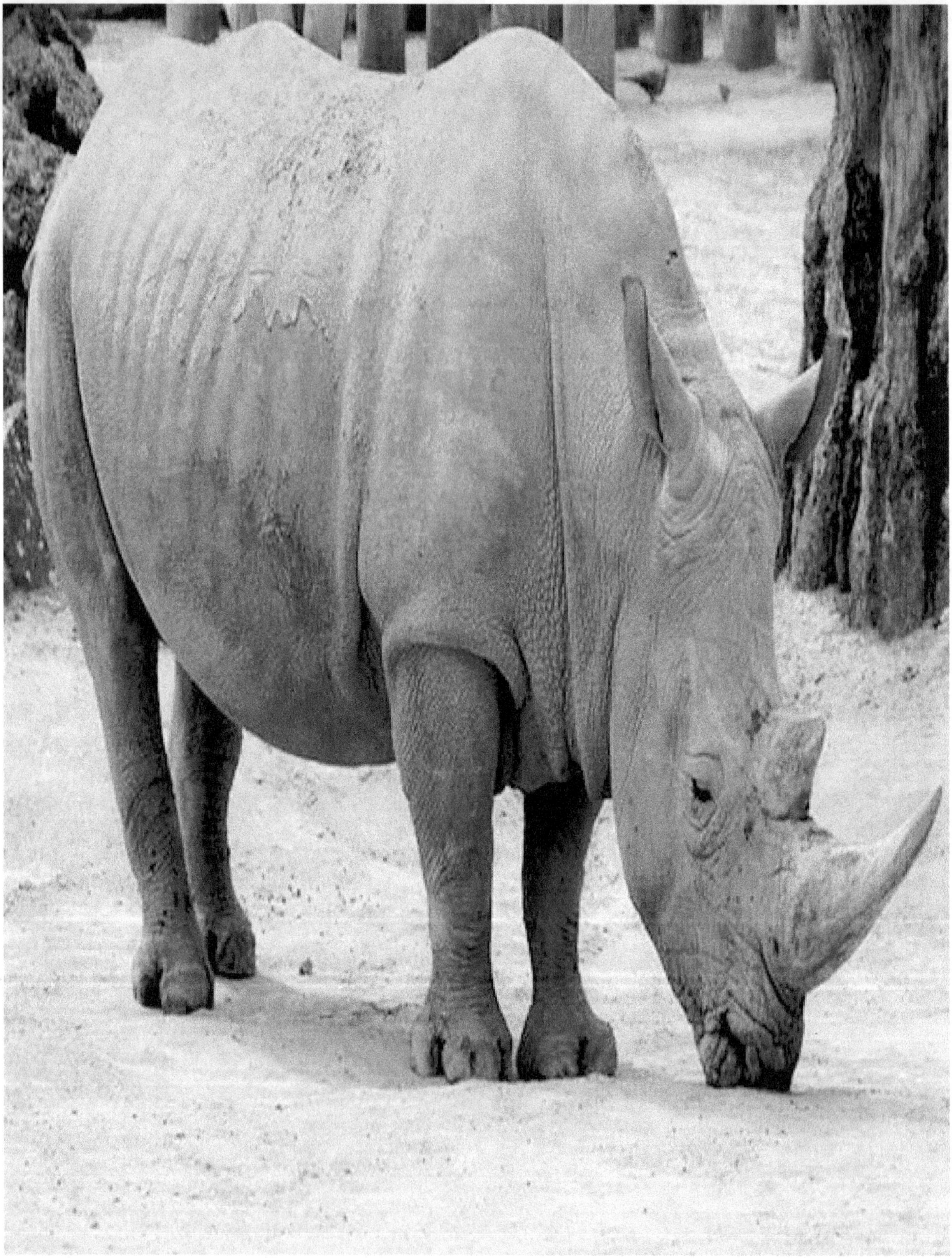

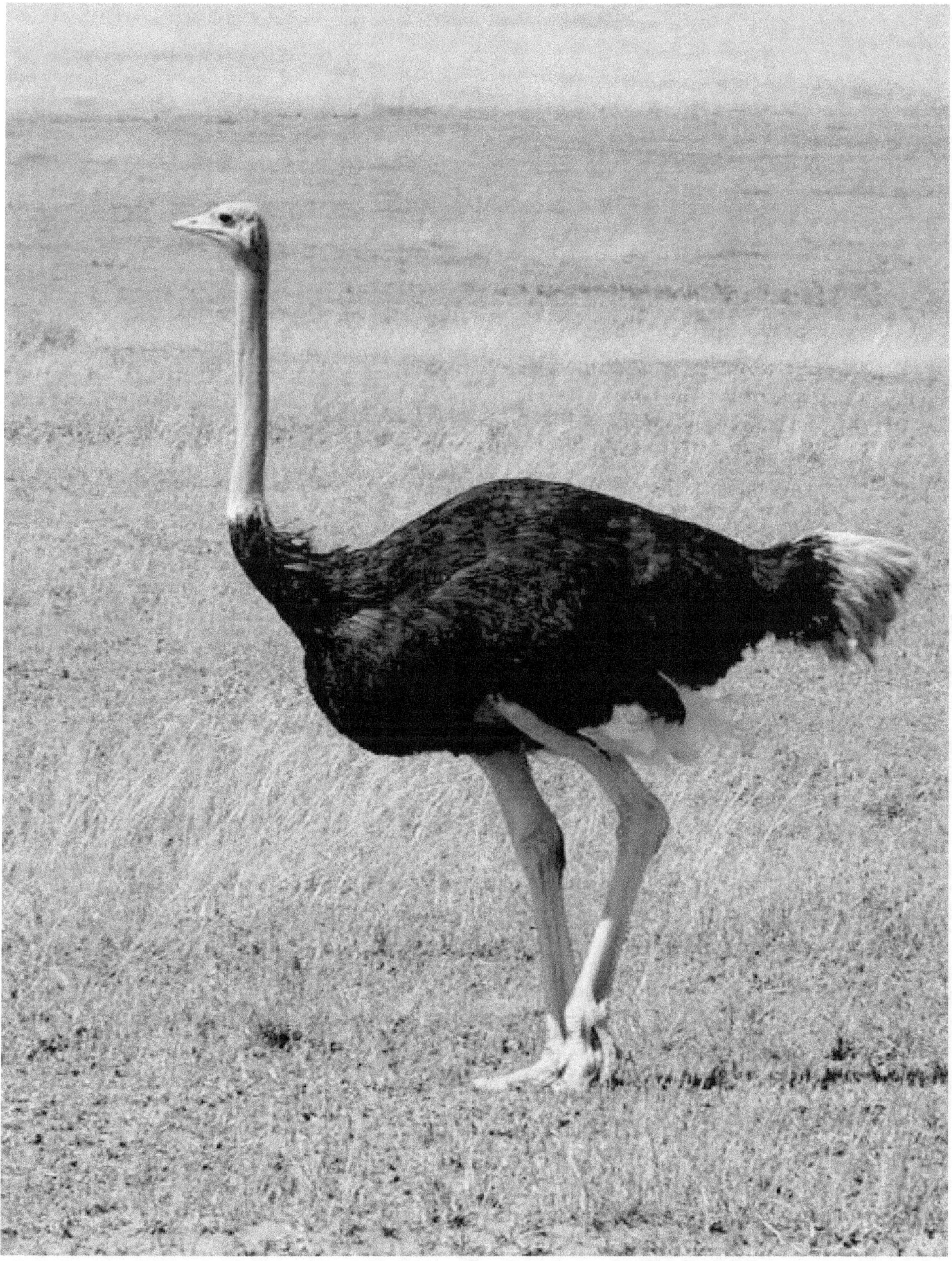

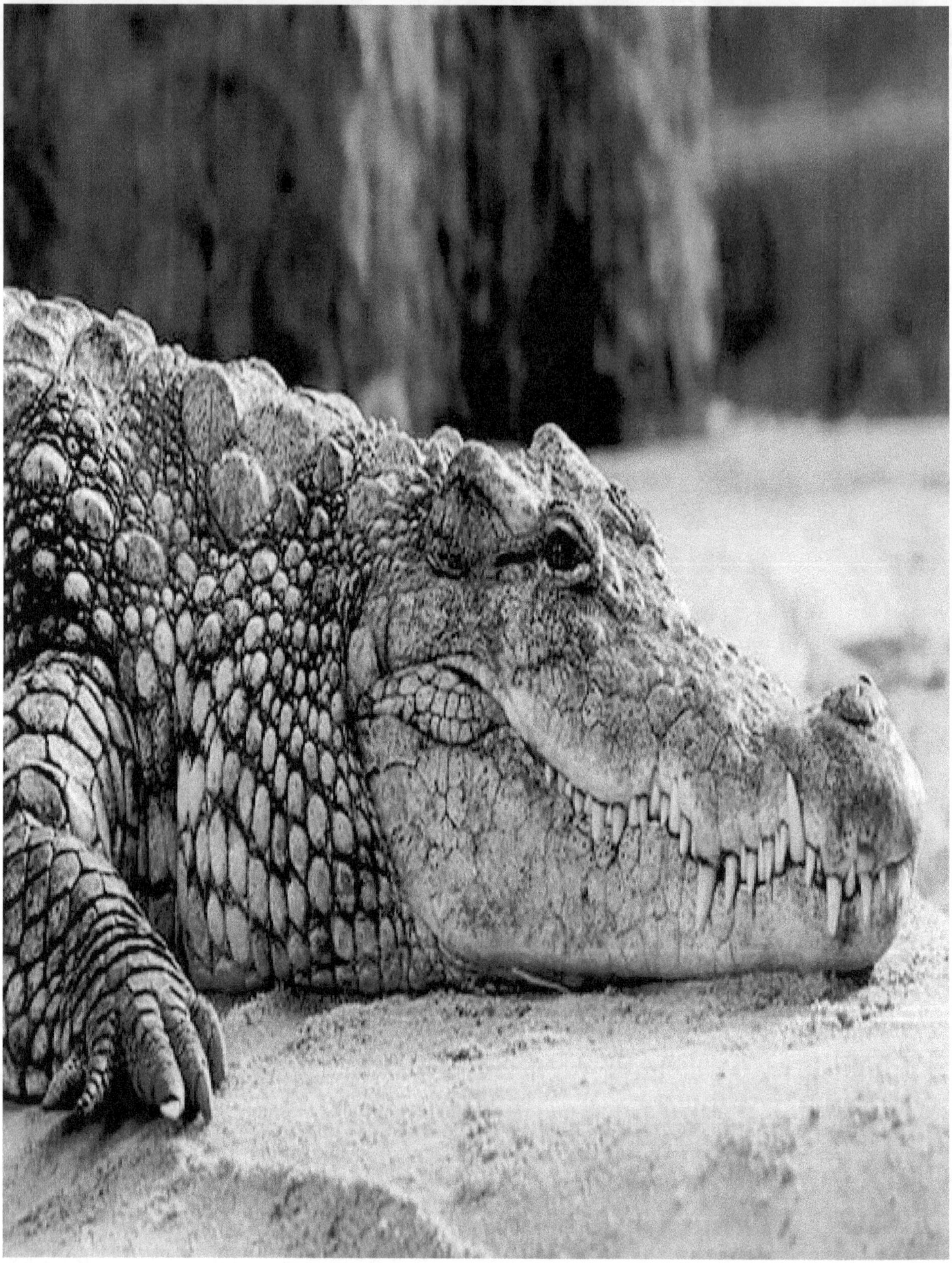

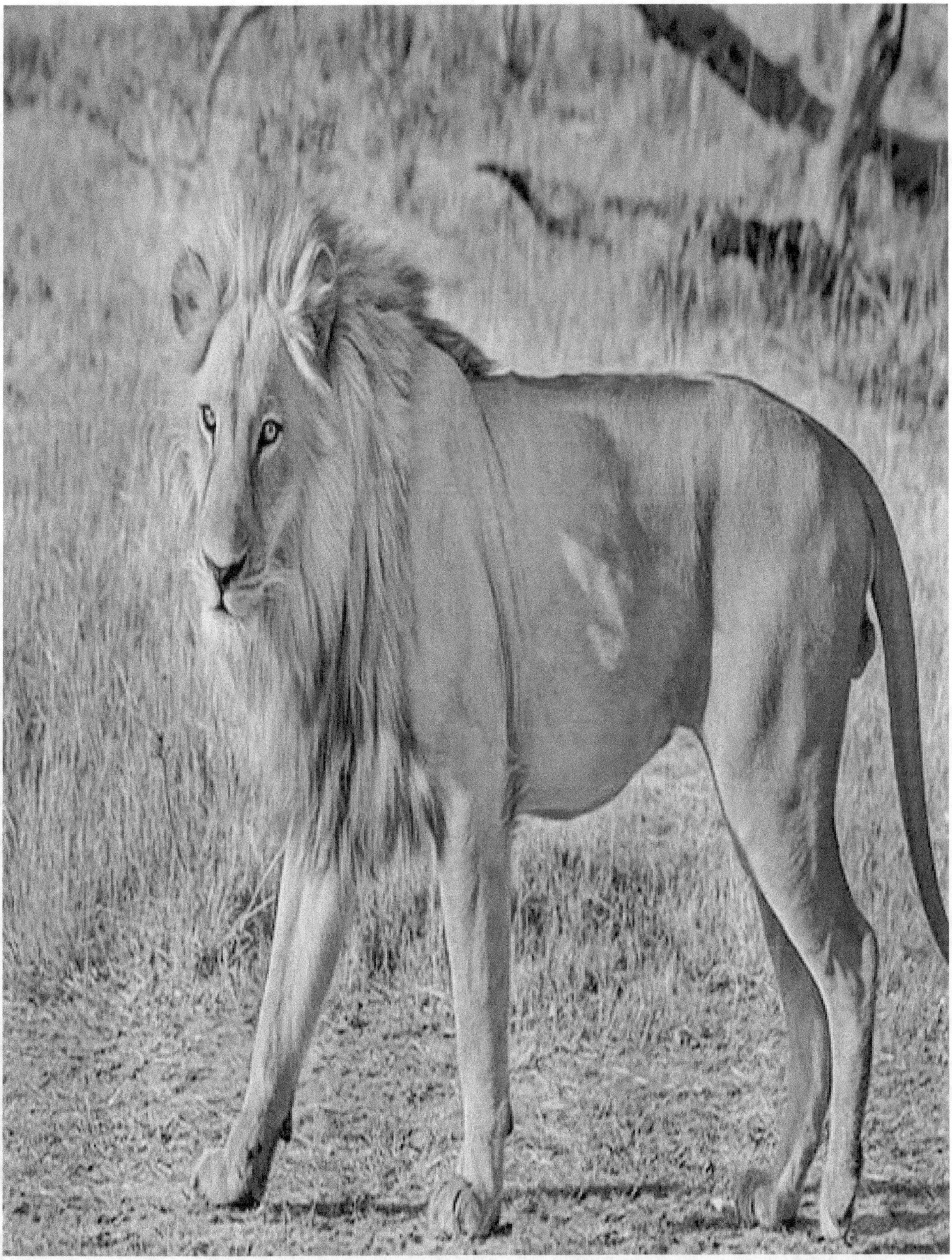

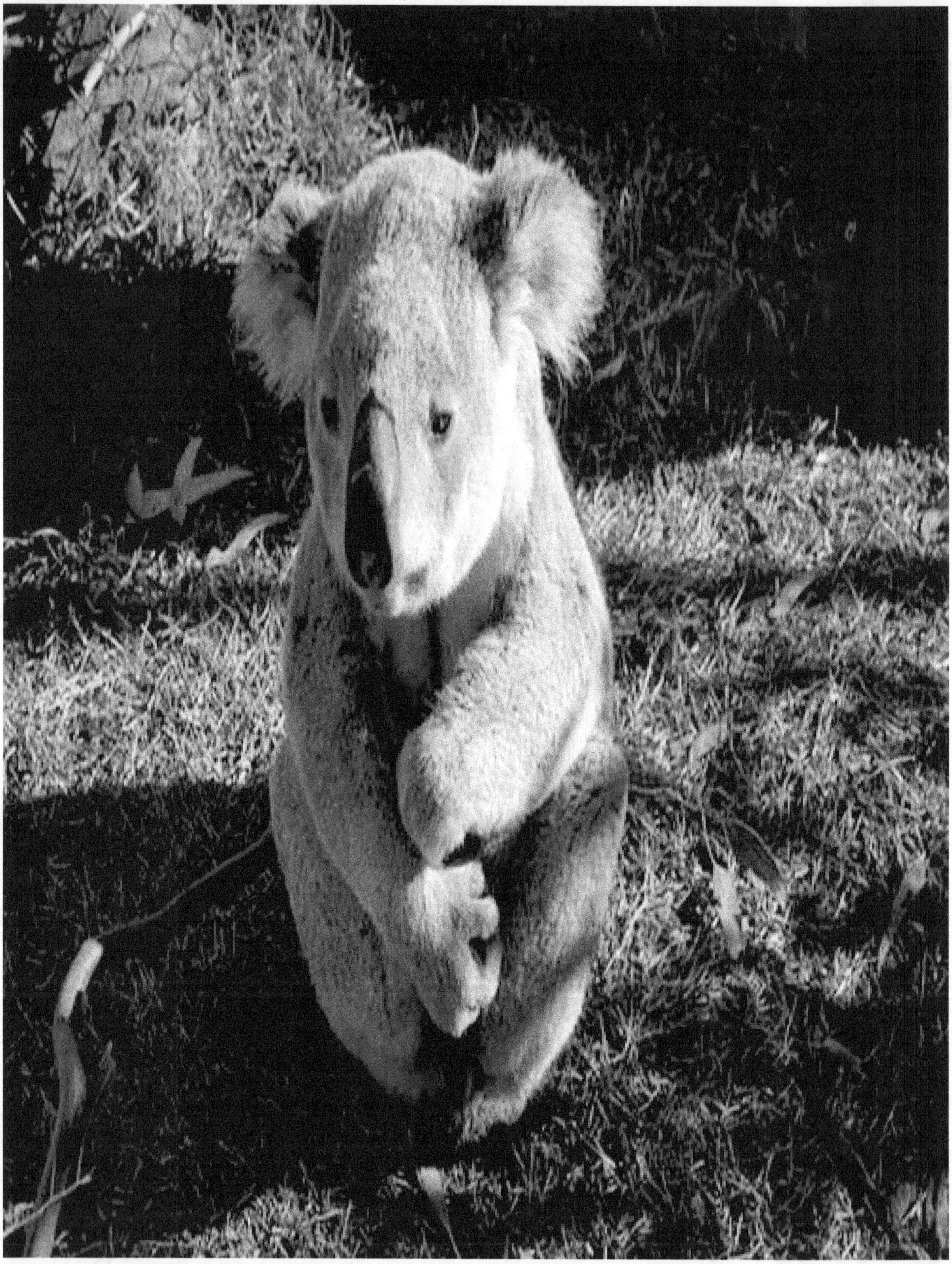

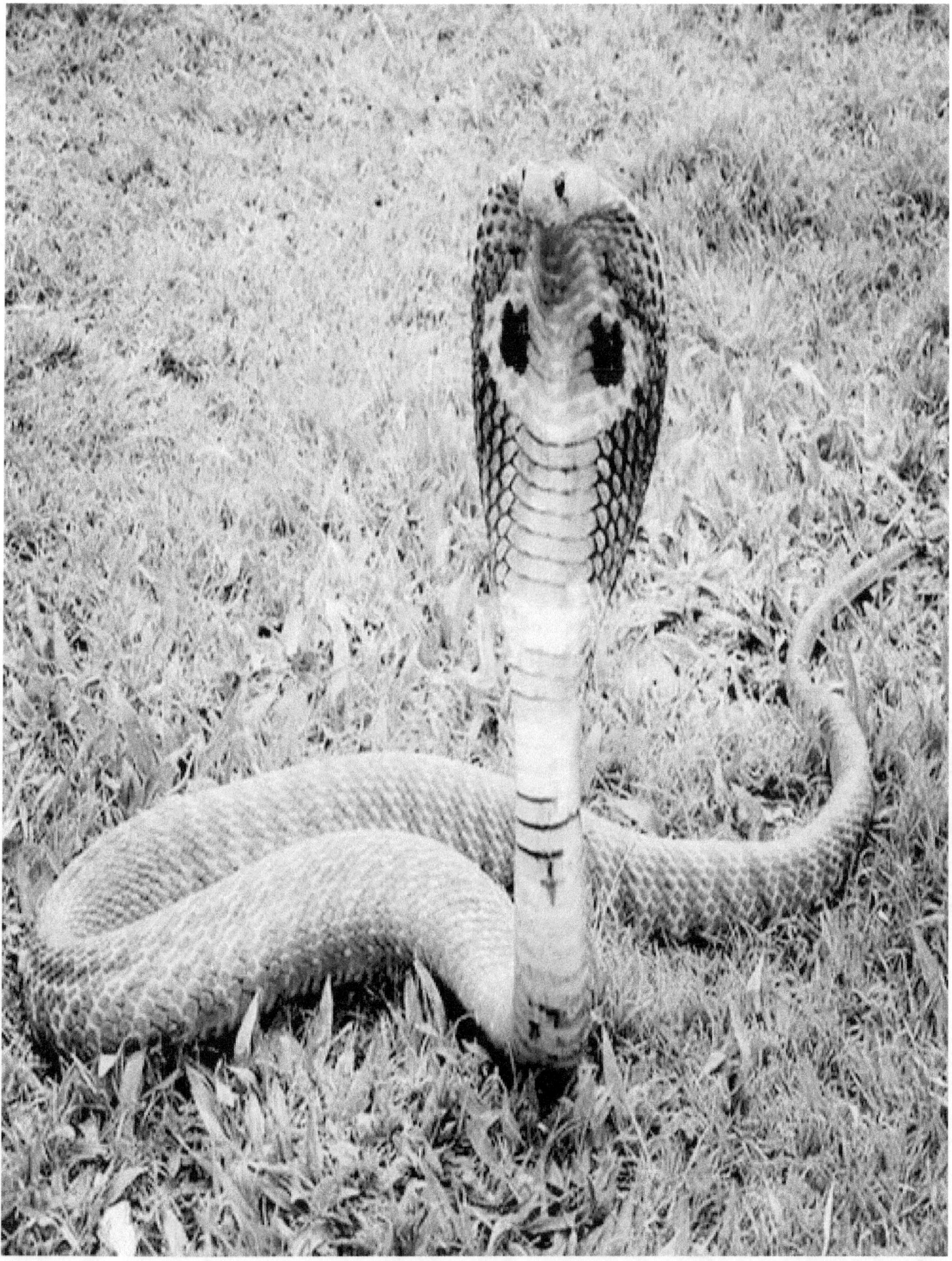

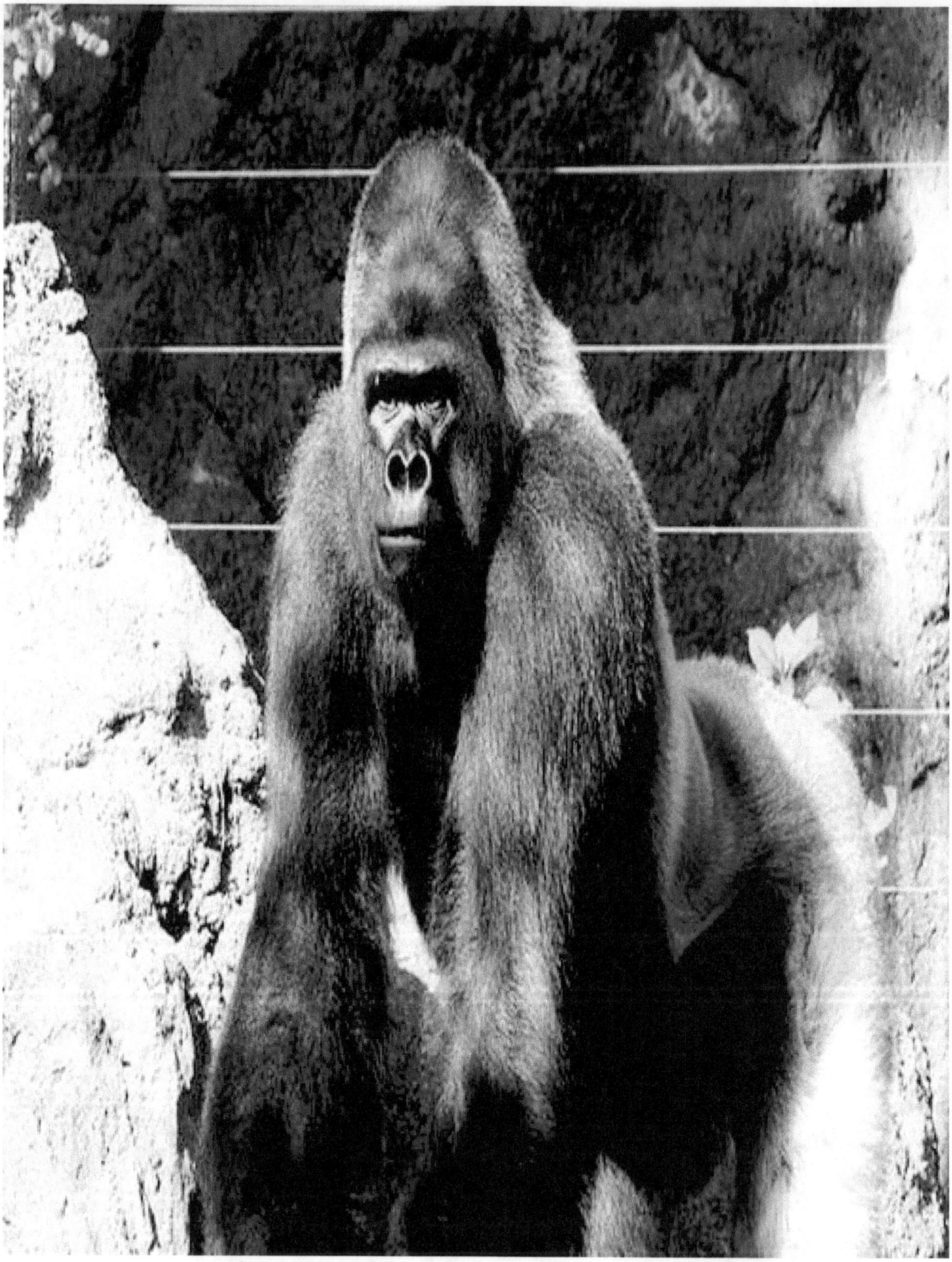

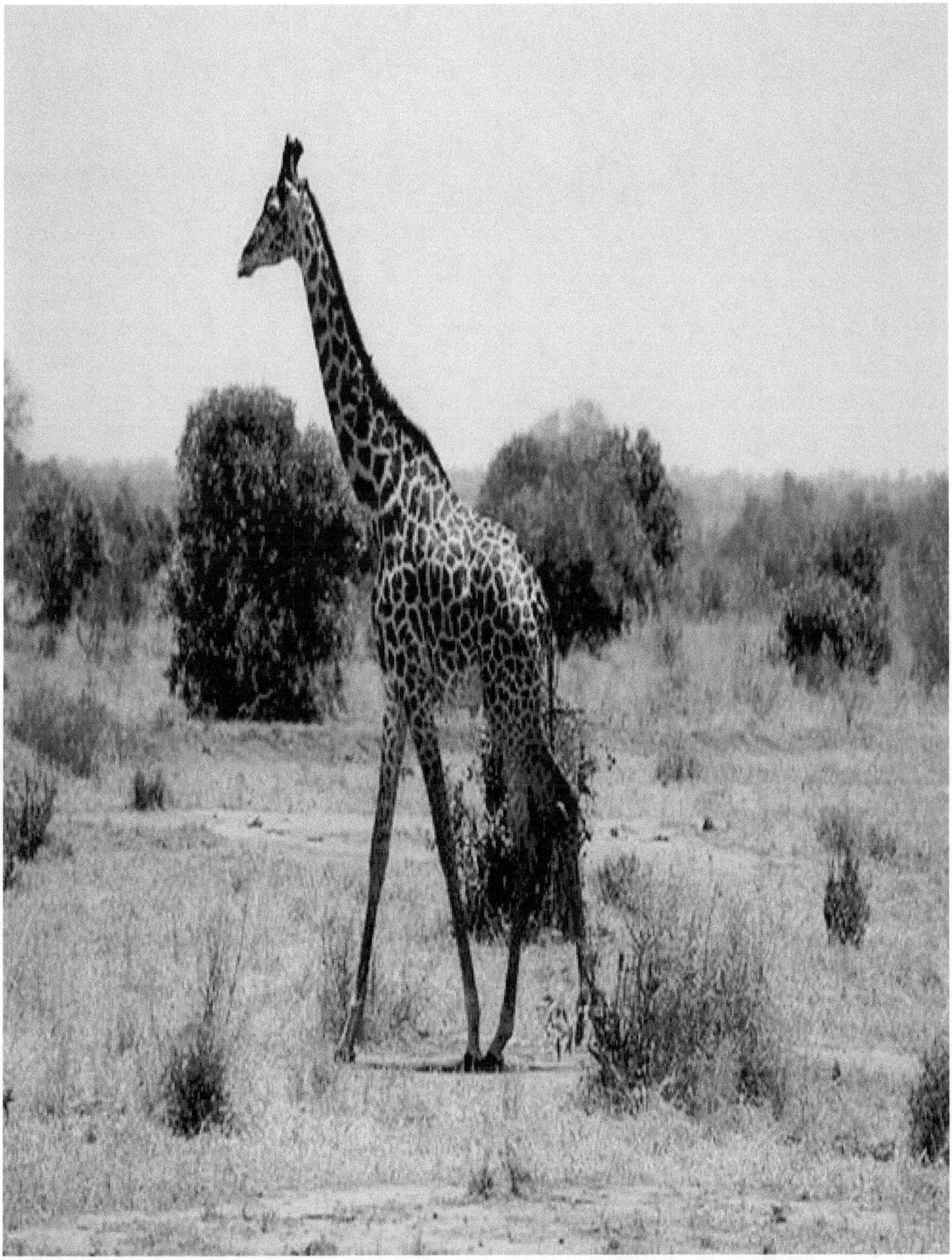